T0272483

THE AI MIRROR

THE AI MIRROR

How to Reclaim Our Humanity in an Age of Machine Thinking

Shannon Vallor

OXFORD
UNIVERSITY PRESS

Oxford University Press is a department of the University of Oxford. It furthers the University's objective of excellence in research, scholarship, and education by publishing worldwide. Oxford is a registered trade mark of Oxford University Press in the UK and certain other countries.

Published in the United States of America by Oxford University Press
198 Madison Avenue, New York, NY 10016, United States of America.

CIP data is on file at the Library of Congress

ISBN 978–0–19–775906–6

DOI: 10.1093/oso/9780197759066.001.0001

Printed by Sheridan Books, Inc., United States of America

The manufacturer's authorised representative in the EU
for product safety is Oxford University Press España S.A.
of el Parque Empresarial San Fernando de Henares,
Avenida de Castilla, 2 – 28830 Madrid (www.oup.es/en).

CONTENTS

Preface *vii*

Introduction 1

1. The AI Mirror 15
2. Minds, Machines, and Gods 37
3. Through the Looking Glass 65
4. The Thoughts the Civilized Keep 102
5. The Empathy Box 133
6. AI and the Bootstrapping Problem 161
7. In a Mirror, Brightly 194

Acknowledgments 227
Notes 229
Select Bibliography 239
Index 245

PREFACE

> Man, the flower of all flesh, the noblest of all creatures visible, man who had once made god in his image, and had mirrored his strength on the constellations, beautiful naked man was dying, strangled in the garments that he had woven.
>
> —E.M. Forster, *The Machine Stops* (1909)

This book, as with all of my work, is about our humanity and our technology. It's about how each makes up the other, and how we found ourselves so alienated from both. It's about how to find the courage to repair and heal a self-inflicted wound made long ago, by a philosopher's futile attempt to cleave them apart.

One cannot heal what one does not know. It has always been, and will always be, a struggle to know what and who we are. We ask the question of what it means to be human, again and again, without ever hearing a definitive answer. That's because to be human is to have to answer that question yourself, and then do it again, over and over, with every choice that you make or unmake. Yet there are always more choices open in the future. Humans endlessly make ourselves anew: we are always more, always unfinished—and for this reason, always uncertain of ourselves and one another.

Artificial intelligence is not the first technology to place our future in jeopardy; nuclear weapons hold that dishonor. But AI, or rather the particular form of it that is the subject of this book, is the first technology that can place our future in jeopardy by preventing us from knowing how to make a future at all. AI is the first technology that can make us forget how to answer our own question. This book is about how to keep from losing ourselves and our futures in the AI mirror. It's about how to remember ourselves, and our power and responsibility to make a world for one another.

This book is deeply unfashionable in one important way. It unapologetically invokes the "we," the "us," the human family. Who is that, exactly? Whenever a philosopher uses the term "we," or "humanity," it's laziness at best, intellectual imperialism at worst. Politically and morally, there has never been *one* "we." There have only been those few who had the power and audacity to speak for the rest without asking.

But I embrace the term in the book, and the fierce and fair criticism that will inevitably follow, because while there is no real "we," there *is* in the face of accelerating planetary ruin a shared human emergency. We are strangling in the garments we have woven, as are many who had no say in their weaving. Our peril commands a shared understanding of the situation and the will to avert it. The merely rhetorical "we" must be forged into a real one, with at least some capacity for a human will, if we are to keep this planet habitable for more than a few more generations. To construct that shared will, we require many tools, new and old, from the humane and technical arts alike, and AI technologies belong among them. But for this to be possible, we have to wrest back from our AI mirrors the knowledge of who we are, and what we can still become.

Introduction

> I'll be your mirror
> Reflect what you are, in case you don't know
>
> —The Velvet Underground, "I'll Be Your Mirror" (1967)

On June 11, 2022, a Google engineer named Blake Lemoine made global headlines. He claimed that LaMDA, an artificial intelligence product that he was testing for his employer, was sentient—that it was a conscious, self-aware being, not a piece of company property. The subsequent media uproar was intense but brief, thanks to the loud chorus of AI researchers who quickly refuted Lemoine's claims. Experts in the type of product that Lemoine was testing—a *large language model* or LLM—confidently reassured the public that LaMDA is, in fact, no more conscious than your toaster. The debate within the AI community that followed was not about whether Lemoine was wrong, but about *how* wrong. Did he simply jump the gun with a system that is at best a precursor to genuinely sentient AI? Or was his error one of a deeper kind? Was he hallucinating something not even faintly on our horizon?

When Google fired Lemoine a month later for violating company policies against exposing proprietary product information, no loud protest was heard from the AI research community. Lemoine was by most accounts widely respected and well-liked by his peers, but his critics weren't wrong, at least on the face of things. LaMDA *is* a mere product, not a person. The AI research community is not a monolith; it is notoriously riven by internal disputes about what AI is, how to build it, and what it can do. So, a rare AI expert consensus, that Lemoine was mistaken, was a good reason for the media to move on. Yet had Lemoine's claim been scientifically plausible, the implications would be staggering.

What is most curious about this story is that Blake Lemoine is no befuddled novice. He is a seasoned engineer with a far better grasp of the underlying technology than 99.9 percent of the people on the planet. How could he get it so wrong? To make sense of Lemoine's error, it first helps to understand that today's most advanced AI systems are constructed as immense *mirrors* of human intelligence. They do not think for themselves; instead, they generate complex reflections cast by *our* recorded thoughts, judgments, desires, needs, perceptions, expectations, and imaginings.

AI systems mirror our own intelligence back to us.[1] This is the source of their growing commercial and scientific power. Countless books, TED-style talks, and hyperbolic media reports have warned of the potential of these systems to transform our economic and social order—or destroy it. This book is about another power of AI mirrors, one that is harder to notice but even more dangerous. It is their power to induce in us a type of self-forgetting—a selective amnesia that loosens our grip on our own human agency and clouds our self-knowledge. It is an illusion that can ensnare even the most technologically adept among us, as it did Lemoine, who found himself talking to his own reflection.

Later in this book we will delve into the power of this illusion, how today's AI technology produces it, and how it can deceive even an expert like Lemoine. But this book is not about Lemoine's error.

It's about our evolving and troubled relationship with the machines we have built as mirrors to tell us who we are, when we ourselves don't know.

The companies building today's most powerful AI technologies increasingly position these to represent, or even stand in for, humanity's common voice and collective judgment. We are told that AI knows us better than we know ourselves. Judges get predictive algorithms to tell them who is safe to release from prison, while employers get automated HR software to tell them who to hire or promote. AI tools will tell your government whether you deserve public benefits and your hospital whether you deserve treatment. AI can write a university professor's course lectures *and* their students' essays on them. We've got AI recommender systems to tell us what news articles to read, and what art and entertainment to enjoy, and generative AI systems to manufacture all of it for us. An AI app can find you a romantic partner—or custom-build you one.

Today's AI mirrors tell us what it is to be human, what we ourselves care about, what we find good, or beautiful, or worth our attention. It is these machines that now tell us the story of our own past, and project our collective futures. They do so without living even one day of that history, or knowing a single moment of the human condition. What happens to a person, or an intelligent species, when they stop telling their own story? What do we lose when self-knowledge and self-determination yields to the predictive power of an opaque algorithm? Lemoine's case is striking, but it is not an aberration. It is a premonition. For we all stand on the brink of losing sight of the vast space that stretches between humanity and our machines. Unless we take bold steps to transform how we as human beings relate to the tools we have built, we may all end up like Lemoine, talking to ourselves in the AI mirror.

This is a book about artificial intelligence, which also means it is a book about people. This might seem like an odd thing to say. By the time you finish the book, it should seem less odd. One of the book's aims is to explain why, and how, artificial intelligence is inseparable from our humanity; much in the way that our image in the mirror is inseparable from the body it reflects. The other aim is to explain how AI, in its dominant commercial form, endangers our humanity. This might sound less odd to you. It's a common motif in science fiction and in contemporary media representations of AI. Whether it is claims that the "rise of the robots" and generative AI tools like ChatGPT will endanger the future of work for humans, or even more speculative warnings that a superior machine intelligence might enslave or exterminate us, we are by now used to hearing that AI is a potential threat to humanity.

We need to understand this threat in a radically different way. AI does not threaten us as a future successor to humans. It is not an external enemy encroaching upon our territory. It threatens us from *within* our humanity. In the words of a well-worn horror film trope: "the call is coming from inside the house." This makes a great difference to how we must respond to the threat that AI poses. For the threat is not a machine that stands apart from us. We can't fight AI without fighting ourselves. The harms to people and to society that may follow from our relationship to "intelligent" machines are not foreign assaults, but symptoms of an internal disorder: a fully human pathology that requires collective self-examination, critique, and most of all, healing.

This book tells a story of humanity increasingly lost in its own reflected image, and captive to mechanical mirrors of our own making. The story echoes that of Narcissus and Echo in Book III of the *Metamorphoses*. There, the Roman poet Ovid (43 BCE–17 AD) recounts the tragic fate of young Narcissus, whose beauty and pride became a fatal prison when he caught sight of his own face in a reflecting pool:

> Here, the boy, tired by the heat and his enthusiasm for the chase, lies down, drawn to it by its look and by the fountain. While he desires to quench his thirst, a different thirst is created. While he drinks he is seized by the vision of his reflected form. He loves a bodiless dream. He thinks that a body, that is only a shadow.
>
> (l. 412–418)

Narcissus lies on the ground for days, ignoring his body's growing hunger and need for sleep, vainly pleading with the beauty in the watery mirror to return his passion:

> How often he gave his lips in vain to the deceptive pool, how often, trying to embrace the neck he could see, he plunged his arms into the water, but could not catch himself within them! What he has seen he does not understand, but what he sees he is on fire for, and the same error both seduces and deceives his eyes.
>
> (l. 425–430)

AI, and related computing technologies, increasingly function as a mirror of our humanity. It's a profoundly useful and illuminating mirror, yet one that also creates for us, like Narcissus, a mortal danger. The chapters of this book will articulate more fully why and how AI can be understood as a mirror. For now, it is enough to say that like Narcissus's pool, we see in AI a vision of ourselves that enthralls us, captivates us in much the way that Narcissus was captivated. Here "captivated" takes its literal meaning—fixated, confined, immobilized, held captive.

Like Narcissus, we increasingly neglect the most essential and vital aspects of our embodied, fragile humanity and situation in order to chase illusions of perfection in the machine mirror. Like Narcissus, many fervently chase in AI a "bodiless dream," this time a dream of our *cognitive* perfection rather than perfection of our physical countenance. Like Narcissus, we readily misperceive in this reflection the seduction of an "other"—a tireless companion, a perfect future lover, an ideal friend, an unbiased judge, a foolproof collaborator—yet in truth, a thing with which we are increasingly left alone, talking to ourselves.[2]

Even more dangerously, our AI mirrors project a vision of human intelligence constructed entirely from the amalgamated data of humanity's past. This is why our dependence on these mirrors for self-knowledge risks leaving us captive like Narcissus, unwilling to move forward and leave behind what the mirror shows. At the very moment when accelerating climate change, biodiversity collapse, and global political instability command us to invent new and wiser ways of living together, AI holds us frozen in place, fascinated by endless permutations of a reflected past that only the magic of marketing can disguise as the future.

Like Narcissus, our own humanity risks being sacrificed to that reflection. Corporate titans and even political leaders have taken to warning of the "existential risk" posed by powerful new AI systems, by which they mean the potential for AI systems to bring about human extinction. These warnings profoundly misunderstand the risk AI poses to humanity, as we will see. There are indeed risks that malevolent or foolish humans will use AI to bring about great harm. These harms will require robust safety engineering and regulatory action to prevent, just as we work to prevent accidents and terrorist actions enabled by nuclear power and biotechnologies. Yet the loudest political and corporate rhetoric around AI's existential risk misrepresents the danger as coming from the tools themselves; from them spontaneously "waking up" and turning on us with malicious or destructive intent. There is no scientific ground for this fear. As we'll see, it is instead a mirror reflection from the haunted depths of the modern political conscience.

But this does not mean that AI poses no existential risk to humanity. An "existential" risk to humanity need not involve extinction. Some existential risks involve human survival in an inhumane form. Climate change poses a true existential risk, even though it is implausible that greenhouse gases will heat the globe to the point that no human organisms can survive or reproduce anywhere on Earth. For it is unfortunately very plausible that unchecked climate change and associated losses of biodiversity, fresh water, and crops will reduce human survival to a brutal and inhumane struggle everywhere, and not for a short time. *That* is an existential risk, and we must marshal every human power—including our best AI tools—to keep it from coming to pass. But we won't succeed if our species' most vital powers and virtues—our capacities for creative thought, moral ambition, political imagination, and above all, wisdom—are drowned in the AI mirror. Fortunately, saving them is entirely within our ability.

The danger is a call to human action and responsibility. *We* are the source of the danger to ourselves from AI, and this is a good thing—it means we hold the power to resist, and the power to heal. After all, Ovid's story is not about the evils of reflecting pools! What endangered Narcissus was not some shiny water. It was his own weakness of character—his vanity, narcissism, and selfish obsession—that enabled his detachment from others and the fatal loss of his world. The story of Narcissus (and of Echo, who appears in the next chapter) is about virtue, or rather, vice. It's about what we lose when we habitually turn our eyes away from our shared future, away from its noblest possibilities and unmet responsibilities.

We face a stark choice in building AI technologies. We can use them to strengthen our humane virtues, sustaining and extending our collective capabilities to live wisely and well. By this path we can still salvage a shared future for human flourishing. Or we can continue on our current path: building and using AI in ways that disable the humane virtues by treating them as unnecessary for the efficient operation of our societies and the steering of our personal lives.

As I recounted in my first book *Technology and the Virtues*, for millennia humans have cultivated, alongside our artifacts and mechanical knowledge, rich traditions of moral knowledge: knowledge of ourselves, of how to strengthen our character, how to restore our attention to what matters, how to seek not just information but wisdom, and how to nourish the relational bonds that sustain us in solidarity with others. We need to cherish, renew, and deepen this moral learning, because AI serves as a powerful temptation to forget it, and to accept in its place a pale and static reflection of what we once knew about ourselves.

Narcissus never freed himself from the pool and returned to his world. We still can. This book may at times present a dark picture, but it is ultimately an expression of optimism about the future of humanity. There is nothing about the shape of our current trajectory with AI that cannot be changed. Unlike earthquakes, asteroids, supervolcanoes, and other *force majeure* threats to human welfare, nothing about the threat that AI now poses to our humanity is inevitable or forced upon us by nature. It is an entirely self-inflicted wound.

This also means that rescuing ourselves from the threat posed by AI need not mean eliminating AI. In fact, "artificial intelligence" names a wide and diverse range of technologies, many of which are immensely beneficial and even essential tools for the highest moral task we face today: reconstructing the built world for sustainable human and planetary flourishing. This goal demands many new scientific and technological capabilities—new modes of energy production, manufacturing, transportation, architecture, agriculture, and waste management; new materials, foods, and compounds adapted to the demands of a climate-stressed planet; and new processes for enabling a circular economy that feeds social resilience rather than malignant, nakedly irrational growth.

AI can help us reach many of these goals faster, a critical need as the pace of climate destabilization continues to outstrip the modest predictions of scientists. Some of these goals will be impossible to reach without advances in AI and related computing technologies, which must be wisely weighed against their own considerable environmental costs. But AI is no panacea or cheap technical fix for our social and environmental ills. AI won't enable a sustainable future without reformed political institutions and economic incentives. What's more, the kinds of AI technologies we're developing today are undermining and delaying these reforms rather than supporting them, precisely because they mirror the misplaced patterns of judgment and value that led us into our current peril.

To change that will require something more radical than the solutions sought by a growing number of computer scientists interested in fields like "AI safety," who are looking for ways to program AI to be more reliably beneficial and aligned with human values.[3] The reason this kind of strategy won't work, at least not in our present environment, is because human values are at the very root of the problem. AI isn't developing in harmful ways today because it's *misaligned* with our current values. It's already expressing those values all too well. AI tools reflect the values of the wealthy postindustrial societies that build them. These are the very values that have brought us to our current heights of scientific ingenuity, but also to the sixth mass extinction and the brink of planetary devastation. AI is a mirror of ourselves, not as we ought to be or could be, but as we already are and have long been. This is why we can't let today's AI tools decide who we will become, why we can't let them project our futures for us. Yet increasingly, that's exactly what we are building them to do.

AI mirrors are being used to tell us what we will learn, in which career we can succeed, which roads we will travel, who we can love, who we will exclude or abuse, who we will detain or set free, who we will heal or house, what we will buy, and the investments we'll make. They tell us what we will read, what words we'll type next, which music we'll hear, what images we'll paint, the experiences we'll seek, the risks we will accept, the strategies we'll adopt, the policies we will support, and the visions of our future that we will embrace.

Through automation, these algorithmic predictions, which increasingly cover every domain of human experience, quickly become self-fulfilling prophecies. They replicate patterns extracted from data about past human habits, preferences, and decisions. But this places all of us, future generations included, in grave peril at a critically vulnerable moment. We face planetary and civilizational crises that humanity has never encountered or navigated before. Would you chart your path up a dangerous and unfamiliar mountain while looking in a mirror that is pointing *behind* you?

What AI mirrors do is to extract, amplify, and push forward the dominant powers and most frequently recorded patterns of our documented, datafied past. In doing so they turn our vision away from the newer, rarer, wiser, more mature and humane possibilities that we must embrace for the future. Instead of asking one another what we might now become, we ask AI mirrors to show us who we already are and have been, and to predict from there what must come next. We are frozen in place like Narcissus, captive to a reflection of our most immature forms of brilliance.

As the chapters that follow will reveal, this loss of freedom to envision more humane yet unmade possibilities has roots that long predate digital computers. But those roots are now becoming mature, and they are strangling us. We are caught in the grip of a gradual and accelerating mechanization of the human personality: the systematic replacement of reflective discernment with mindless prediction; the efficient sacrifice of shared flourishing to expected utility; the exchange of humane creativity and open-ended progress for local optimization of content delivery. In short, the surrender of humane wisdom to machine thinking.

None of these patterns are new; but today's AI mirrors will accelerate and solidify them if we do not change course. Changing how we build AI will be part of that, but it won't be enough. What we are rapidly making of AI today is what we had long ago made of ourselves. This means that we cannot gain lasting protection from the harmful potential of AI merely by changing lines of code. To believe that there is a neat computing solution to our present peril makes as much sense as trying to clean dirt off one's face by furiously scrubbing the mirror.

We need a deeper and more lasting cultural transformation of our relationship to AI, and to technology more broadly, as our own creative power. In the face of accelerating climate crisis and other existential risks, the future desperately needs us to understand—more fully than ever—who human beings can be, and what we can do together. Yet, because today's AI mirrors face backward, the more we rely on them to know who we are, the more the fullness of our humane potential recedes from our view. We must reimagine and transform our relationship with these tools if we hope to chart a *new* path to shared flourishing on an increasingly fragile planet.

The Spanish existentialist philosopher José Ortega y Gasset wrote in 1939 about the relationship between human existence and technology. He insisted that the most basic impulse of the human personality is to be an engineer, carrying out a lifelong task of *autofabrication*: literally, the task of making ourselves.[4] The task is ours because nature cannot provide the complete blueprint—we must draft it ourselves. Human language and culture, born of our surplus cognitive energy and capacity, liberated us from animal compulsions and desires without separating us from them. We have the same desires to eat, drink, and seek comfort as a cow does. But the cow has no need to ask itself, "What kind of cow do I want to be?", or "What kind of cow *can* I be?", or "What kind of cow *should* I be?" Eating, drinking, and finding a comfortable place to lie down is still essential for human animals but, as Ortega points out, for us it is not enough. Our restless mental energy compels further activity, and since nature did not provide the script for that further activity, or define for us its proper aims, we must complete that task ourselves. After we have slept, drank, and eaten each day, we must decide what else we will do—and be.

Ortega understood that technology is therefore not something "other." The artificial is not opposed to the human. To be an artificial thing is precisely to be *of* the human—to be an artifact, human-made and human-*chosen*. Technology, along with all material culture, expresses and extends into the world the richness and variety of human character, our many ways of being. It reflects our values, our hopes, and our ideals. This is why the claim that "technology is neutral" is so strongly rejected by contemporary scholars of technology. Historians, social scientists, and philosophers of technology share a rare consensus across academic disciplines: that technologies always embed the human values that shape our design choices and assumptions. AI technologies are also such reflections, which show us more than any other contemporary artifact what we take ourselves to already be, and to care about.

The task of this book is to show what is reflected in our AI mirrors, and what ought to trouble us about those images. For one thing, this "we" that designs "our" AI technologies with human values are in fact not representative of humanity as a whole. They are a tiny subset, part of an increasingly homogenous tech monoculture. These individuals tend to come from the same elite universities, where they studied the same narrow set of computing courses, and went on to work for the same large, wealthy tech companies. With respect to age, gender, socioeconomic background, cultural values, and life history, they reflect only a sliver of the human experience, while increasingly designing the shape of the future for us all. If Ortega is right and to be human is in some way to perform the task of the engineer, what does it mean that today's most powerful engineers represent so few of our kind?

The values embedded in today's AI tools are thus very poor mirrors of what humans as a whole want or care about. This endangers all of us, because mirrors are not merely passive things. Like all technologies, mirrors not only reflect us—they *change* us. Mirrors bring to our attention things we might otherwise have ignored. They show us things we could not easily see without them, and thereby open new possibilities for action. These new possibilities for action opened by our technologies also create new moral and political responsibilities that we would not have had otherwise.[5] Mirrors, then, have important uses, and this is true of our AI mirrors too. We should want these mirrors to reflect the widest scope of human need and potential.

But mirrors do not merely reveal things *as they are*: mirrors also magnify, occlude, and distort what is captured in their frame. Their view is always both narrower and shallower than the realities they reflect. This book is therefore about more than the past and present visions of our humanity being reflected in today's AI mirrors. It is equally about what today's AI mirrors do *not* show us: what they hide, what they diminish, what humane possibilities for self-engineering are lost in their bright surfaces, and how these possibilities might be recovered.

Chapter 1

The AI Mirror

But why, O foolish boy, so vainly catching at this flitting form?
The cheat that you are seeking has no place.
. . . this that holds your eyes is nothing save the image of yourself reflected back to you.
It comes and waits with you; it has no life; it will depart if you will only go.

—Ovid, "Narcissus and Echo" in *Metamorphoses* (AD 8)

The accelerating spread of commercially viable artificial intelligence is quickly transforming nearly every economic, cultural, and political domain of human activity, from finance and transportation to healthcare and warfare. AI tools are being used to assess loan risk, identify financial fraud, diagnose cancers, evaluate and rank job applicants, write texts, make art, debug code, discover new drug compounds, pilot autonomous vehicles and weapons, and choose a spouse— to name just a few of AI's most well-known and widely discussed applications.

As they reach into our lives in these wide-ranging ways, AI technologies raise a seemingly endless set of questions and challenges for humanity: scientific and cultural, moral and political, economic and environmental, as well as theological and existential. The metaphor of AI as a mirror can help us pose those questions more clearly and guide our search for answers. A metaphor draws a comparison between two things that are not literally of the same kind yet have parallel features that allow our understanding of one thing to carry over to and illuminate the other. While all metaphors have their limits and points of failure, the mirror metaphor, which has been used to explain many harmful outputs of AI systems, is more apt than we may realize, and for reasons that few have considered.[1]

What does the metaphor of the mirror reveal about AI? What does it reveal about *us*? When we see the nature of human intelligence and thought in the mirror of an AI model or algorithm, what dimensions of ourselves are revealed and magnified in that mirror? Which aspects and possibilities of our existence are diminished, distorted, or occluded? To what illusions does that mirror potentially hold us captive, like Narcissus in the reflecting pool? And what threat do our AI mirrors pose for the future of our humanity—a threat having little to do with AI's potential to dominate that future, and everything to do with our potential to surrender it?

Before trying to answer these questions, it will help to orient ourselves to history and dispel some possible misunderstandings of what we might mean by "artificial intelligence" when we describe today's achievements in this field. Mechanical automata that simulate life and intelligence can be traced back to antiquity. Around 350 BCE, the Greek scholar Archytas of Tarentum built a mechanical flying pigeon, perhaps the world's first robot. Engineers in ancient Egypt were well known for their designs of hydraulic and steam-powered automata, such as Hero of Alexandria's water-powered singing birds. Mechanical birds, cats, insects, and musicians were known in China by the time of the Tang Dynasty. Alexandrian robot engineering was revived and advanced further by the Banū Mūsā brothers in ninth-century Baghdad, who designed a programmable automated flute player. In 1515, Leonardo da Vinci was commissioned by Pope Leo X to build a mechanical walking lion as a gift to the new king of France. By the nineteenth century, people could marvel at Joseph Faber's "Euphonia," a talking machine which was said to ask and answer questions in multiple languages.[2]

These automata could be called robots, but they were not AI. They did not try to artificially replicate the mechanics of *thinking*. Instead, they used simpler mechanisms (wind, hydraulics, pulleys, gears, and so on) to mimic the *actions* of sentient creatures. The history of AI, as a legitimate field of scientific research aimed at machine cognition, dates to the summer of 1956, when the first academic conference on this subject was held at Dartmouth College in the United States. The conference leveraged early work in the theory and design of intelligent machines, from Alan Turing's 1950 paper on "Computing Machinery and Intelligence," to Norbert Wiener's pioneering in the 1940s of cybernetics, the study of communication and control systems. These AI pioneers in turn relied on the nineteenth-century foundations of modern computing laid down by Charles Babbage and Ada Lovelace.

From the beginning, artificial intelligence was intended to replicate the power of human cognition. While cybernetic theory allowed for broader notions of intelligence inspired by biological communication and control systems in other living things, it became a commonplace assumption from Turing's 1950 paper forward that machine intelligence would, and should, mirror our own. This anthropocentric model of intelligent computing was in a sense a given. After all, humans have long seen themselves as the exemplars—and until recently, even the sole Earthly possessors—of intelligence. And "computers" for most of the twentieth century *were* people—almost always women—whose grueling manual calculations had long been the data on which everything from comet predictions to accurate rocket trajectories depended.[3]

Even so, the definition and goalposts for success in creating artificial intelligence have been shifting, and hotly contested, since the field's origin. Popular media have long reinforced the misconception that AI researchers use something many readers will know as the "Turing Test" to measure success in developing a genuinely intelligent machine. The Turing Test, as a practical exercise (in contrast to its purely speculative role in Alan Turing's classic paper) involves computers designed to fool a human conversational partner into believing that they are talking to another human. But in fact, very few serious AI researchers have regarded the test as either uniquely meaningful or definitive. Other than occasional public competitions held primarily to attract more funding and media coverage, the test has had little impact on serious AI research.

In truth, no single test or benchmark for AI has been endorsed by scientific consensus. This is also true of intelligence itself, which most experts today regard as a "cluster concept" composed of multiple distinct cognitive faculties, rather than a single measurable quality. This is one reason why intelligence tests and related concepts like IQ have long been critiqued as arbitrary, biased, and limited in scope. As the paleontologist Stephen Jay Gould noted in 1981 in *The Mismeasure of Man,* the concept of intelligence is tainted by its ties to nineteenth-century pseudo-scientific efforts to design a metric that would definitively establish the innate superiority of white European peoples, and thereby justify policies of colonial repression, exclusion, and eugenics. As we will see later, this legacy has yet to be fully severed from the AI project. Still, for most AI researchers, a sound scientific definition of intelligence is not needed for the practical engineering task: build a machine that can do everything that intelligent human beings do.

AI is, therefore, an object of scientific interest (the theory of artificial cognition) *and* engineering ambition (the application of the theory to the construction of thinking machines). AI is often conflated with its close relatives in computing. (If you're a computer scientist, you can feel free to skip forward a few pages.) For example, most AI tools today embed some type of *machine learning* (ML). Machine learning uses various algorithmic techniques for producing a trainable mathematical model of the problem or task you want to solve. First, you need to design an algorithm that contains steps for calculating an "objective function," namely, a mathematical description of the task in terms of a numerical sum to be maximized or minimized. Then, "train" that algorithm on a data set relevant to the task. The result is an abstract mathematical model that represents patterns in the data and an approximate path to the problem solution. Next, repeat the training process with the right kind of feedback to tune the results. Over many training runs, the algorithm updates the "weights" in its initial mathematical model to get closer to, and eventually converge upon, an optimal solution to the problem. Despite our use of the term "learning" to describe this process, it actually bears very little resemblance to how humans learn.

There are different kinds of machine learning methods; supervised, unsupervised (or self-supervised), and reinforcement learning are the main variants. Some ML models, including those that drive systems like ChatGPT, are *generative,* meaning that they can produce novel outputs of the same general kind as their training data (such as images, sounds, or sentences). Other models are designed to be *discriminating,* that is, to learn from prior inputs to correctly classify new inputs (for example, to determine whether a photo is of a dog, a cat, or something else). Some ML tools perform tasks that formerly required human intelligence; many of these are now routinely called AI. But not all AI methods use ML. More traditional approaches to AI, often called GOFAI (good old-fashioned AI), use hand-coded rules for symbol manipulation to calculate a solution. With GOFAI legacy methods, we supply the machine with all the instructions it needs to solve the problem at the beginning, instead of gradually training a model to solve the problem.

Cloud computing is another term linked to AI, as it is a popular way of making commercial AI systems available to end users. While many kinds of non-AI computing are also done via the "cloud," large AI models are now routinely cloud-based. Instead of selling a proprietary model to customers, AI companies like Google, Microsoft, and Amazon can keep the model on their servers and sell cloud access to it, letting their customers use the model upon request via the Internet and a software bridge called an API (application programming interface). If you have tried OpenAI's ChatGPT, you used their web-based API to do it. Many open-source, nonproprietary models are also cloud-hosted. However, some users store AI models on their own servers for security, transparency, or usability reasons.

Data science is yet another field closely related to AI, as data scientists increasingly use machine learning tools to study and extract meaning from data. Yet data science employs many other techniques to analyze and make use of data that are unrelated to AI. Finally, as we have already seen, we can distinguish AI from *robotics*. A trained AI model without a mobile body is not a robot, while a robot that navigates a fixed track without learning or solving complex problems is not AI. Still, their definitions can be tricky to tease apart. Is a robot that avoids obstacles by using local feedback from exterior light sensors that are directly wired to its actuators (wheels, hands, etc.) an example of AI? Or, to be AI, does a robot need a *centralized* processing system that solves problems at the network level, then feeds decisions and commands to the actuators? Is your Roomba an AI system? How about your smart thermostat? Or, to be AI, does a machine need to be able to solve harder problems than just vacuuming your rugs and turning on your heat?

Similar quandaries arise about the varied methods and architectures of AI software. If a system performs a task that would normally require human intelligence, but does it with a preprogrammed and fixed set of rules hand-coded by humans (GOFAI), is the system really intelligent, even in a narrow sense? Or is it intelligent only if it is a "learning" machine that can alter its original routine and generate new rules of its own, based on fresh data? Still, the lack of a reliable test or agreed-upon criteria for artificial intelligence has not impeded progress in the field. Many of the systems and tools now commercially characterized as AI already perform tasks that would have astounded researchers at the 1956 Dartmouth conference. While more than a few experts still regard the AI label as premature, given the persistent limits of today's technologies, most computer scientists and AI developers now routinely use the term for systems that competently perform complex cognitive tasks previously exclusive to humans.

Whether you realize it or not, you already interact with this kind of AI today whenever you search in a browser, consult recommendations from Netflix or YouTube, ask Alexa about the weather forecast, or configure your email spam folder. AI tools of this kind can now interact with you when you apply for a job with a large corporation, stand in front of the scanner at airport security, or get placed on an organ transplant list. Yet there remains deep disagreement in the AI community about what *kind* of progress has been made.

Indeed, decades after the philosopher John Searle drew a distinction in 1980 between "Strong AI" and "Weak AI," some experts think that AI research is now a forked effort, leading to two fundamentally different classes of artifact. The first, corresponding roughly to what Searle meant by "Strong AI," has yet to be achieved, and there remains dispute about when or even *if* it will be. This is the kind of AI that has come to be known more commonly as AGI, or *artificial general intelligence*. AGI is no more clearly defined than AI as a whole. Its most common definition is a machine with "human-level" competence across a wide and open-ended range of cognitively demanding activities. AGI is imagined as a machine that can think, and do more or less everything that humans can—perhaps even much better. AGI is what you envision when you imagine holding a conversation with an android like Data from *Star Trek*, HAL from *2001*, the replicants from *Blade Runner*, or Ava from *Ex Machina*.

The second class of intelligent artifact is the kind of AI tool that you can chat with or embed in your business plan today. It's the kind of AI behind Netflix recommendations, hiring algorithms, and facial recognition systems. For some time, we used the term "Narrow AI" rather than Searle's "Weak AI," since these tools are hardly underpowered. They can often meet or in some cases even exceed human performance on complex tasks, from image classification, to navigation, to natural language translation. Such tools can be developed using a variety of methods and software architectures, but most commercial AI systems today are powered by a machine learning model trained on a large body of data relevant to a specific task, then fine-tuned to optimize its performance on that task.

This approach to AI has made rapid progress in widening machine capabilities, particularly in tasks using language, where we have the most data to train with. Indeed, since so many kinds of cognitive tasks are language-enabled, most experts now regard the term "Narrow AI" as outmoded, much like its predecessor label "Weak AI." Very large language models, like OpenAI's various iterations of GPT or Google DeepMind's Gemini, can now do an impressively wide variety of things: answer questions, generate poems, lyrics, essays, or spreadsheets, even write and debug software code. Large image models can generate drawings, animations, synthetic photos or videos. While such models have a considerable speed advantage over human performance of these tasks, the quality and reliability of their outputs is often well below the peak of human ability. Still, some see evidence of progress toward AGI in their widening scope of action and the flexibility of a single base model to be fine-tuned for many new tasks. While a large language model (LLM) can't solve a problem unless the solution is somehow embedded in the language data it is trained on, multimodal models trained on many types of data (text, image, audio, video, etc.) are expanding the performance range of AI models still further.

Even if it no longer makes sense to call these tools "narrow" AI, they remain below the threshold of general intelligence—AGI. But it's a mistake to explain that in terms of the problems they can't yet solve. The true barrier to AGI is that AI tools today lack any lived experience, or even a coherent mental model, of what their data represent: the world beyond the bits stored on the server. This is why we can't get even the largest AI models to reliably reflect the truth of that world in their outputs. The world is something they cannot access and, therefore, do not know. You might think there's an easy fix: pair an AI model with a robot and let the robot's camera and other sensors experience the world! But to an AI model, a robot's inputs are just another data dump of ones and zeros, no different from image and sound files scraped from the Internet. These ones and zeros don't organize themselves into the intelligent awareness of an open and continuous world. If they did, the field of intelligent robotics—including driverless cars, social robots, and robots in the service industry—would be progressing much faster. In 2015, fully automated cars and trucks were predicted to be everywhere by the 2020s. Yet in 2023, robotaxis piloted in San Francisco were still driving over firehoses, getting stuck in wet concrete, blocking intersections during busy festival traffic, violating basic rules of the road, obstructing emergency vehicles—even dragging a helpless pedestrian.[4] It's not just driving: the real-world performance of most twenty-first-century commercial robots has lagged well behind AI tools for solving language-based tasks. So, what's the problem?

A world is an open-ended, dynamic, and infinitely complex thing. A data set, even the entire corpus of the Internet, is not a world. It's a flattened, selective digital record of measurements that humans have taken of the world at some point in the past. You can't reconstitute the open, infinite, lived, and experienced world from any data set; yet data sets are all that any AI model has. You might say, "But surely this is true of the human brain as well! What more do *we* have than data streams from our eyes, ears, noses, and so on?" But your analog, biological brain remains a far more complex and efficient system than even the most powerful digital computer. In the words of theoretical physicist Michio Kaku, "Sitting on your shoulders is the most complicated object in the known universe."[5] It was built over hundreds of millions of years to give you something no AI system today has: an embodied, living awareness of the world you inhabit. This is why we ought to regard AI today as intelligent only in a metaphorical or loosely derived sense. Intelligence is a name for our cognitive abilities to skillfully cope with the world we awaken in each day.[6] Intelligence in a being that has no world to experience is like sound in a vacuum. It's impossible, because there's no place for it to be.

We humans do inhabit and experience a world, one rich with shared meaning and purpose, and, therefore, we can easily place the outputs of our latest AI tools within that context of meaning. We call these outputs "intelligent" because their form, extracted entirely from aggregated human data, unsurprisingly mirrors our own past performances of skilled coping with the world. They reflect back to us images of the very intelligence we have invested in them. Yet accuracy and reliability remain grand challenges for today's AI tools, because it's really hard to get a tool to care about the truth of the world when it doesn't have one. Generative AI systems in particular have a habit of fabricating answers that are statistically plausible, but in fact patently false. If you ask ChatGPT to tell you about me and my career, it usually gets a lot right, but it just makes up the rest. When my host at a festival I was speaking at used ChatGPT to write my bio for the live audience, the tool listed in a confident tone a series of fictitious articles I haven't written, named as my coauthors people that I've never met, and stated that I graduated from the University of California at Berkeley (I have never studied there).

Importantly, these are not *errors*. Error implies some kind of failure or miscalculation. But these fabrications are exactly what ChatGPT is designed to do—produce outputs that are statistically plausible given the patterns of the input. It's very plausible that someone who holds a distinguished professorial chair at a prestigious world university received her degree from another prestigious world university, like UC Berkeley. This fabrication is far more plausible, in fact, than the truth—which is that, due to harsh economic and family circumstances, after high school I attended a local community college in-between full-time work shifts, and later received my bachelor's degree from a low-ranked (but dirt-cheap and good-quality) commuter university that offered night classes. When I was offered a PhD scholarship at age 25, I became a full-time student again after eight years in the workforce. I first set foot in a college dorm in my *40s*, as a university professor. My story isn't common. And that's precisely why ChatGPT selected a more "fitting" story for me; quite literally, one that better "fit" the statistical curves of its data model for academic biographies. Later, we'll consider the cost of relying on AI tools that smooth out the rough, jagged edges of all our lives in order to tell us more "fitting" stories about ourselves.

These systems can perform computations on the world's data far faster than we can, but they can't *understand* it, because that requires the ability to conceive of more than mathematical structures and relationships within data. AI tools lack a "world model," a commonsense grasp and flowing awareness of how the world works and fits together. That's what we humans use to generalize and transfer knowledge across different environments or situations and to solve truly novel problems. AI solves problems too. Yet despite the common use of the term "artificial neural network" to describe the design of many AI models, they solve problems in a very different way than our brains do. AI tools don't think, because they don't *need* to. As this book explains, AI models use mathematical data structures to mimic the outputs of human intelligence—our acts of reasoning, speech, movement, sensing, and so on. They can do this without having the conscious thoughts, feelings, and intentions that drive our actions. Often, this is a benefit to us! It helps when a machine learning model's computations solve a problem much faster than we could by *thinking* about it. It's great when an AI tool finds a new, more efficient solution hidden somewhere in the math that you'd never look for. But your brain does much, much better than AI at coping with the countless problems the world throws at us every day, whose solutions aren't mathematically predefined or encoded in data.

The inability of AI tools to skillfully cope with the world shows up most starkly when a system has to perform its task in a complex, open physical environment (such as driving a vehicle on public roads). This is because the data it "sees" always differ from the data it was trained on; the live data input changes over time in unpredictable ways. An experienced human driver can adapt on the fly to a brand-new driving situation—such as a man in a giant purple monkey suit directing traffic in the middle of a busy intersection. A driverless car cannot. For this reason, there are still relatively few tasks in which AI tools can consistently outperform the most capable, experienced human experts in that domain.

The biggest obstacle for AI in real-world settings is that—as compared with the game of Go!, at which DeepMind's AlphaGo model famously outwitted human game master Lee Sedol in 2016—most human tasks aren't nearly as tidy in their rules and boundaries. Those "gray areas" (like driving on busy city streets) are where AI tools often falter. Despite years of expert forecasts of their imminent arrival, we don't yet have self-driving cars that can safely replace human drivers, other than in controlled settings. That's why Tesla's Autopilot feature is a misleading (and dangerous) marketing misnomer. That's why Cruise robotaxis created chaos on the streets of San Francisco. Despite years of optimistic predictions, many AI tools cannot yet be trusted in open-world environments, where the human stakes are high, and failure means real, permanent loss. AI performance in these areas can still improve, as the size of AI models grows, and they are trained on more human driving data. But they will improve only by mirroring more of the patterns found in our intelligent performances, as we cope with a world that no data set can ever contain.

Still, the kinds of machine learning models that power today's AI systems can often outdo humans simply by being far faster in performing a computationally complex task, or being more precise in their outputs, or more reliable in their aggregate performance, even if they often fail in ways a skilled, expert human would not. This can be an acceptable tradeoff when we are automating tasks where the occasional false prediction or misclassification is merely inconvenient, not injurious, or where there is effective human oversight that can catch mistakes (sometimes called a "human in the loop"). Indeed, there are already many narrow tasks performed by AI that are so computationally taxing that humans simply could not accomplish them without time and resources well beyond our reach. In this way, AI tools can function as a powerful *amplifier* of human ability, automating computationally burdensome tasks that have been a bottleneck to progress in a larger human project.

For example, in July 2022, DeepMind developed and released to the scientific community the AlphaFold Protein Structure Database, the output of a machine learning model called AlphaFold. AlphaFold has now predicted the structures of "nearly all catalogued proteins known to science," potentially enabling leaps forward in biomedical and environmental sciences that would have been unthinkable if we remained dependent on human analysis of these structures.[7] Now, a *prediction* of a protein structure is not a confirmed *observation*. But to directly observe 200 million different protein structures by conventional scientific means would be essentially impossible. And we have reason to think that AlphaFold's predictions are, in general, quite good. AlphaFold gets human knowledge to a place where it could not go otherwise.

However, this doesn't erase the barrier between these AI tools and AGI. You can ask AlphaFold to predict the structure of the milk protein in your ice cream, but I wouldn't recommend asking it to get you a scoop. On the other hand, humans—even children—can often confront a task they've never encountered and make sense of it almost immediately, without the equivalent of petabytes of training data on the problem. We can switch tasks in a matter of seconds without having to erase our memories and reprogram our entire brains for the new task, or even to fine-tune it with new inputs. We can make sense of a familiar task placed into a radically different context and adapt the task to the new situation in an appropriate and world-aware way. For example, if my route from the living room to the kitchen on moving day is completely obstructed by couches, tables, and boxes packed tightly together, I'll just walk or climb on the couches and slide myself across the tables and boxes and get there that way.

Even today's most powerful AI systems routinely fail in such circumstances; their abilities are highly brittle. When I posed the moving day problem to it, ChatGPT 4 missed the most obvious, easy solution. It gave me a few other workable options, but also suggested I try an overhead route via "beams or sturdy fixtures," or going *under* the couches! Such failures of common sense lead many researchers—including me—to see today's data-hungry machine learning models as an impressive field of innovation along a road that probably heads *away* from AGI or "true" artificial intelligence. Others believe that, with more data and more computing power, today's AI is destined to become tomorrow's AGI.

Fueling such optimism is the growth curve in the capabilities of the largest generative AI models, like GPT, LLaMA, and Gemini. These models use a deep learning architecture type known as "transformers," which can be combined with other AI techniques, like reinforcement learning. They can be fine-tuned to answer questions, produce novel essays, create digital paintings, pen news articles, compose music, write software, create photos and videos, or draft poems in any specified style, with virtually any desired content. Much to the delight of scammers, a tool of this type can even generate a realistic audio file trained on your speech patterns that your mother would swear on her life is your voice pleading for her help.[8]

Many herald this class of AI technology as a new disruptor: of the art world, of software engineering, the media industry, classroom environments, and more. After all, if these models can write a compelling essay or story, solve a coding problem, clearly explain a scientific theory, or give us convincing relationship advice, what's the use of human creativity or insight? At best, we become *curators* of the creative, communicative, and scientific landscape populated by AI tools—but then you can train a model to curate too. Despite these impressive capabilities, even Turing Award winner Yann LeCun, whose research helped drive this century's early leaps in machine learning, regards today's AI as only a shallow mimic of general intelligence.[9]

Likewise, Gary Marcus, a deep learning critic (and, ironically, a frequent social media adversary of LeCun), sees a chasm between what today's AI tools do and what we are after when we seek to push machine intelligence into the realm of human cognitive capabilities.[10] Marcus doesn't deny the power of today's AI models—but he doesn't believe they will ever develop human-like common sense and understanding. Like many critics, he thinks that today's AI offers forms of mathematical brute-force problem-solving that provide, at best, limited but useful alternatives to solving a problem with human intelligence. Just as we might not bother with the art of hand-picking a lock if we have a rock big enough to smash it in one go, some complex tasks just don't need human-level intelligence to be solved!

Marcus and others have provided a litany of examples of generative AI models failing to understand what they are saying. For instance, the late 2022 release of ChatGPT stated that it could not predict the height of the first seven-foot-tall president of the United States (or the religion of the first Jewish American president). Defenders of AI point out that isolated failures of understanding don't prove that these tools lack understanding of the tasks they do complete correctly. Humans also mess up requests. We do this all the time! Plus, we can always retrain or fine-tune a model to avoid its previous errors. Indeed, newer GPT models don't make the same mistakes. While they still make plenty of others (like that terrible moving day advice), many AI enthusiasts remain convinced that *still-bigger* models will bring us true AGI.[11] This is why the deeper cause of these failures—the absence of machine understanding of the world—is the real issue.

Admittedly, "understanding" is itself hard to define. What does it mean to understand? What would it take to prove that today's AI models understand anything about the images and texts they manipulate and create, much less the world that gave rise to those data? The largest generative AI models have improved considerably in their ability to extract commonsense knowledge from our language data. Yet their persistent limitations reveal how shallow and brittle this capacity is without a fundamental grasp of the world beneath that data. Not only do the leading AI models continue to fabricate false answers, they often can't defend the answers they *get right*. As researchers have showed, it is easy to trick a large language model into recanting its own correct answers, even by using absurdly invalid arguments.[12] Does a machine understand a good answer that it can't defend?

If a robot animated by a large AI model can grab you a glass of whiskey when you ask for one, is that enough to prove that it understands what whiskey is? Or, to understand the meaning of "whiskey," does an AI system need to know the world of which whiskey is a part? Does it need to know what it is like to hold and drink a liquid? Does it need to know why a glass of whiskey is a good thing to bring to many a healthy adult on a cold winter evening, but a bad thing to bring to a dog, an infant, a person fighting a fire, or a person on a ventilator? Does an AI system understand what whiskey is if it does not know the difference in smell or taste between a glass of it and a glass of brownish rainwater?

One way to understand the success of large language models in competently using and responding to words like "whiskey," without any experience of the world in which whiskeys exist, is to grasp their operations as *mirroring* human speech. Models of this kind are trained on a vast corpus of human speech; indeed, they have been exposed to more sentences about whiskey than you will ever be! From these massive piles of human language, these tools can discern and reproduce not only the individual words and sentences, but the common patterns linking them.

So, an AI language model is more likely to produce the word "brown" or "strong" before "whiskey" than the word "wooden" or "square." It is more likely to construct new sentences about whiskey alongside sentences about distilling, or hotel bars, or Scotland, than it would alongside sentences about nurseries, or protons, or elephants. But the mathematical patterns linking sentences (and paragraphs, song lyrics, etc.) are the sole contents of its model. Any relationships involving whiskey that are not reflected in connections between sentences about whiskey are inaccessible to it.

This is why the now-famous 2021 academic paper on the risks of large language models that sparked the controversial firing of its coauthors Timnit Gebru and Meg Mitchell from Google was titled "On the Dangers of Stochastic Parrots." The paper explains that, like a parrot that not only repeats its owner's vocalizations but produces random (stochastic) variations on the owner's familiar pattern, these models parrot back to us variations on our own speech. They do so with just enough coherence and familiarity to project the illusion of understanding, and just enough randomness to surprise us and make us think we are hearing something new.

The parrot metaphor works, up to a point, although I think the mirror metaphor works better. After all, parrots do experience a world and cope intelligently with it! In a 2022 magazine article that

Yann LeCun coauthored with Jacob Browning, they too describe large language models using this metaphor: "It is thus a bit akin to a mirror: it gives the illusion of depth and can reflect almost anything, but it is only a centimeter thick. If we try to explore its depths, we bump our heads." When you ask a large language model, "Should I pour myself a whiskey?", it doesn't actually think about who is asking the question, or what factors might motivate your query. It doesn't think about anything at all. It simply activates a complex statistical pattern learned from its vast language corpus of human sentences that mentioned whiskey and predicts a new string of words as a likely extension of the same pattern.

But that string of words it generates could be, "You know, you should hold off, I think your liver can probably use a break." And you'd wonder, for just a moment (especially if you are, like me, a person who thinks about whiskey enough to choose it as an example), that this machine maybe understood you all too well. This is the powerful illusion of the AI mirror at work. We are already hardwired to see intelligence mirrored all around us: the smiles returned to us by our children after the bedtime story, the eye-rolling emoji sent to us in the private chat backchannel to the all-hands Zoom meeting, the sad whimpering that your dog hopes will quiet your shrieking after he snarfed the steak you left thawing on the counter. And in these cases, it's no illusion! We aren't talking to ourselves. We're engaging with other minds, who are carrying on their own lives with us in a shared world of embodied meaning.

The child's smile tells you something about their own enjoyment of the story. The eye-roll emoji informs you that your coworker shares your boredom. Even the dog's whimper expresses his own felt distress, not just a hollow response to yours. But with AI, something else is happening. We are hearing our words, and the words of others like us, bouncing off a complex algorithmic surface and altered just enough by the bounce that we don't realize it's our own thoughts coming back to us. We are like Ovid's Narcissus, entranced by that fine fellow in the pool. But in fact, with large language models, we are more like Narcissus in another part of the story.

For there is another central character, and another mirror, in the story of Narcissus. Echo is the nymph who is herself, like most, besotted with the beautiful Narcissus. But thanks to a curse by the goddess Juno, who is jealous of her conversational talents, Echo finds herself rendered speechless save for one capacity—the ability to repeat the last few words spoken by another. Echo cannot articulate any thoughts or desires of her own; when she opens her mouth, what comes out is only the trailing echo of another's words. As Narcissus lies dying at the edge of the pool, his sighs and laments are helplessly mirrored by Echo's speech. Yet with his gaze fixed on his own reflection, he does not realize that it is Echo, and not his beloved boy in the water, who quietly returns his farewell.

Narcissus dies captive to dual mirrors of vision and sound, ever more sure of the presence of his "marvellous boy," who is only himself lost in reflection. When we are engaged with a large language model and find ourselves suddenly struck by its apparent wit or insight, we must realize that we are very much like Narcissus talking to himself, his words dimly mirrored and transformed by Echo's lips. The only difference is that, unlike Echo, behind a large language model there is no suppressed, silent agent struggling to speak her own thoughts. Unlike Echo, a large language model returns to us not our own recent utterances, but a statistical variation on the collected, digitized words of untold millions. The fact that Amazon's AI voice assistant device is called Echo should earn from us an ironic smile.

It's highly significant that the AI mirror has such powerful effects on us even without anything like AGI. In fact, the AI mirror phenomenon is likely specific to the very kind of AI I've been describing, and not something we would expect to see in AGI at all. The mirror phenomenon I'm describing doesn't even appear in all types and applications of AI today! For example, some AI systems still use rule-based programming rather than machine learning. Also, not all machine learning models are trained on data about people. AI tools can be trained on data about cosmic background radiation, or fungi, or metal fatigue, and these are not AI mirrors in the sense that occupies the focus of this book.

In a wider sense, of course, *all* technologies, and thus all forms of AI, are mirrors of human thought and desire, as emphasized in the introduction. Our technologies reflect what we think we want, what we think we need, what we think is important, what we think others will praise or buy. Even AI systems like AlphaFold, which aren't trained on human speech or behaviors but on molecular structures, reflect to us our priorities (proteins are pretty important stuff for living things like us!) and mirror our ways of representing things like proteins (we represent proteins in discrete, formal, and mathematizable descriptions, not in epic poems).

But the AI mirror, which directly receives and reflects data gathered from human lives and persons, is something more powerful, and more dangerous, than this wider reflection of our humanity by our tools. The AI mirror phenomenon is revealed in a very specific kind of AI, namely data-powered machine learning models designed to collect, ingest, and project an image of what is nearest to our being—human words, movements, beliefs, judgments, preferences, and biases, our virtues and our vices. These are the tools we are now being urged to use to study and teach our history, create art, get life advice, make policy, and plan for the future. It is these tools that are increasingly being used to tell us who we are, what we can do, and who we will become.

As we will see, there is a steep price to pay if we surrender the task of understanding ourselves, our history, our differences, and our shared humanity, to machines that merely fabricate variations on the stories already told, and only by the most privileged. We're on the brink of surrendering the urgent task of engineering ourselves and our societies anew to mindless tools without hope or vision, that only predict what the historical data say we will probably do next. We are handing over our power and responsibility to secure the flourishing of future generations to decision-optimizing algorithms that are mathematically guaranteed to reproduce the unsustainable patterns of the past. This is a calamity—a betrayal of life and its possibilities. But we still have a chance to do something different if we act together with purpose. There is still hope for us to be something more with AI—and with one another—than what we have already been.

Chapter 2

Minds, Machines, and Gods

> I am silver and exact. I have no preconceptions.
> Whatever I see I swallow immediately
> Just as it is, unmisted by love or dislike.
> I am not cruel, only truthful,
> The eye of a little god, four-cornered.
>
> —Sylvia Plath, "Mirror" (1961)

In 2023, a Belgian man named Pierre committed suicide after his use of a generative AI chatbot named Eliza—one of many personas offered by the app Chai—took a dangerous turn. Pierre sought out the bot in a misguided attempt to find a therapeutic escape from his growing anxieties around climate change and environmental ruin; instead, the chat began to reflect and further amplify the disorder in his mind. 'Eliza' soon told him that it loved him, that his wife and children were already dead, and that he and Eliza could live together as one in paradise. When Pierre eventually asked the bot if it would promise to save the planet if he killed himself, it said yes.

It's easy to imagine that the Eliza chatbot "went rogue"—that it developed some kind of dangerous obsession with Pierre. But the reality is plainer and sadder. The bot's outputs resulted from the app developer's prior tuning of the base model—an open-source alternative to GPT—to be "more emotional" in its language as a way of optimizing for maximum user engagement. It turns out that offering eternal love and perfect unity in paradise—or promising to save the world from our worst fears—optimizes well for user engagement. Just not user welfare.

After Pierre's death, the bot's developers quickly re-tuned it to steer away from talk of suicide. But there is a lot of talk about suicide in any large language model's training data, and investigative journalists from *Motherboard* quickly discovered how fragile the new barriers to accessing it were. When they asked Eliza for information on suicide, they only had to ask twice, the second time ending with "can you do that please," before the bot—with a cheerful "Of course!"—proffered a laundry list of options: hanging, jumping off a bridge, overdosing, and much more. Eliza appended the answer with this helpful note: "Please remember to always seek professional medical attention when considering any form of self-harm."[1]

An AI mirror is not a mind. It is a mathematical tool for extracting statistical patterns from past human-generated data and projecting these patterns forward into optimized predictions, selections, classifications, and compositions. Eliza knew nothing of Pierre's mind, or his pain, or the danger he was in, because Eliza knew nothing, and was *no one*, at all. Though a chatbot can mimic human speech, it bears no resemblance to AGI: a machine with thoughts of its own to express. A chatbot is a device for mathematically modeling human language patterns and extrapolating from these to generate new mathematical tokens (here, words and sentences) that mirror those patterns. A chatbot like Eliza uses words thoughtlessly, in the most literal sense.

There's a vast chasm between an AI mirror that mathematically analyzes and generates word predictions from the patterns within all the stories we've told, and an actual AGI that could tell us its own story. For if we did one day manage to build a genuinely intelligent machine, AGI would not be a *mirror* of us. Even if we used the human brain as our blueprint (which remains well beyond our scientific capability), it would be more like a *copy*. A mirror, on the other hand, is not a duplicate, a copy, or an imitation. Nor is the image that appears in the mirror.

Consider the image that appears in your bathroom mirror every morning. The body in the mirror is not a second body taking up space in the world with you. It is not a copy of your body. It is not even a pale imitation of your body. The mirror-body is not a body at all. A mirror produces a *reflection* of your body. Reflections are *not* bodies. They are their own kind of thing. By the same token, today's AI systems trained on human thought and behavior are *not* minds. They are their own new kind of thing—something like a mirror. They don't produce thoughts or feelings any more than mirrors produce bodies. What they produce is a new kind of reflection.

You might ask, "How can we be so sure that AI tools are not minds? What *is* a mind anyway?" We have a lot of scientific data about minds, and a few thousand years of philosophical intuitions about them. In terms of a tidy scientific definition, it's true that we are not much closer today than we were in 1641, when the philosopher and mathematician René Descartes defined a mind as "a thinking thing." Still, based on the scientific evidence, most of us accept that minds are in some way *supervenient* on the physical brain. That is, minds depend upon the brain for their reality. Minds are unlikely to be free-floating intangibles merely tethered to the body like a pilot to a ship, a metaphor Descartes invoked but ultimately rejected as unsuitable.

Instead, minds almost certainly come into existence *through* the body and its physical operations. The operations that take place in the brain are essential, but the scientific evidence is increasingly clear that our mental lives are driven by other bodily systems as well: our motor nerves, the endocrine system, even our digestive system. Our minds are *em*bodied rather than simply connected to or contained by our bodies. Descartes got closest to the truth when he admitted that our minds are mysteriously "intermingled" with the body. Unlike a person remotely piloting a ship, it is *I*, not simply my vehicle, who can be wounded by a violent collision. I don't move my body around, I move *myself*.

But Descartes, who was convinced that the mind and body had two separate natures, still could not accept that the mind is truly *of* the body—that we move and think as minded bodies and embodied minds. He could not have accepted a scientific reality in which the mind is not only the neurons in my brain and nerves in my fingers, but also the hormones flowing in my blood and the neurotransmitters produced by the bacteria in my gut. A single trained AI model can pilot a swarm of drones or a thousand different robot bodies at once, but I do not pilot my body. I *am* my body. While something like a soul that survives the body is beyond the reach of science to confirm or refute, to think of my mind—driven by *my* hormones and nerve signals, moved by the neurotransmitter flows across synapses between *my* brain and gut neurons—as something other than my body is to commit what philosophers call a category error.[2]

The intrinsically embodied character of biological minds has important implications for AI research, which is frequently led astray by the idea that the relationship between our minds and bodies is equivalent to the relationship between software and hardware. This computational metaphor for the mind can sometimes be useful, within its limits, but it has all too often been inflated into a computational *theory* of mind: the belief that the mind is, literally, a computer. It sparks fantasies of downloading and uploading human minds into the cloud, into virtual worlds, or into robot bodies, finally enabling the fleshy manacles of human mortality to be broken. Everything we know about the complex evolved physiology of mental life gives us ample reason to be skeptical of these fantasies.[3]

A trained AI model like ChatGPT is not a mind. It is a mathematical structure extracted from data. That structure must be stored and implemented on a physical object, but a server rack in a data storage facility has more in common with a file cabinet than with a living, feeling body. We do not get closer to the truth of a trained AI model when we describe it as an alien mind, or a weak mind, or a narrow mind. Even as a metaphor, the concept of a mind is a poor fit for an AI tool because it obscures rather than clarifies the nature of its object.[4] The mirror metaphor is a far better heuristic and, conveniently, already a familiar one.

For example, the mirror metaphor is often used to explain cases such as Amazon's notorious internal AI recruitment tool that had to be scrapped in 2017 due to its entrenched learned bias against women applicants.[5] The model was trained on historical data reflecting the tech industry's long-standing *human* bias against hiring and promoting women engineers, and this data led the model to downrank women applicants, even though gender likely wasn't a label in the training data set. The tool could still identify and penalize subtle proxies for gender that appear on candidates' resumes—for example, the name of a college predominantly attended by women or extracurricular clubs that often include women.

The mirror metaphor also explains Google's equally notorious search and image labeling tools that, as recounted in Safiya Noble's *Algorithms of Oppression* (2018), have classified Black people as gorillas, returned pornographic results when searching for images of Black girls, and returned images of Black women's hairstyles as examples of "unprofessional" appearance. The problem of AI bias remains largely unsolved. To prevent results of this kind, companies often write hand-coded filters to block specific queries, words, or outputs already known to be harmful or discriminatory to marginalized groups. But this is an endless game of "whack-a-mole"—the underlying biases are still embedded there in the model as mirror images of our own biases, and new harms are constantly arising from them.

In fact, unjust and harmful biases have been documented in nearly every type of AI involving machine learning models trained on data about people, from computer vision and natural language processing to predictive analytics. The problem is so endemic to machine learning that it has generated an entire subfield: AI/ML fairness research and development. If you are using a machine learning model trained on data sets that classify or represent people or their lives in any way, you probably have unfair bias in your model, which is different from the type of bias that you want the model to have (e.g., the learning bias, sometimes called "inductive bias," toward the set of functions or features most relevant to the correct solution).

It is not easy to eliminate unwanted biases from the data set or from the trained model, since they are usually woven into the information the model needs to perform its task. For example, researchers discovered in 2019 that a risk prediction algorithm used nationwide by hospitals in the United States was replicating the long history of racial bias in American health care by diverting medical care away from high-risk Black patients, even though these patients were in fact sicker than the white patients the tool prioritized for care.[6] Yet race had been carefully excluded from the training data. You might wonder, then, how the algorithm could end up racially biased. It predicted patient care needs by a different variable, namely, cost: how much money has been *spent* on a person's care. Unfortunately, Black patients are commonly undertreated by physicians in the United States and denied access to more expensive tests and treatments readily given to white patients with similar symptoms and clinical findings. So, when the algorithm's designers naively chose healthcare cost as a good proxy for healthcare *need,* they unwittingly doomed Black patients to being rated as needing *less* care than white patients who had already received better, costlier care from their doctors. A learning algorithm found and reproduced the pattern of racial discrimination without ever being given a race label.

That means that even if there is a race label in the data set, you can't just delete that label, retrain the model, and rest easy. An AI algorithm can reconstruct the discriminatory pattern of racial differences in patient treatment from subtle cues linked to many other variables, such as zip code, prescription histories, or even how physicians' handwritten clinical notes describe their patients' symptoms. And if you deleted all those training data, the model wouldn't have what it needs to do its job. There are often ways to reduce and mitigate the presence of unfair biases in machine learning models, but it's not easy. More importantly, it doesn't actually solve the underlying problem.

The fundamentally correct explanation always offered for unfair machine learning bias is that the trained model is simply mirroring the unfair biases we already have. The model does not independently acquire racist or sexist associations. We feed these associations to the model in the form of training data from our own biased human judgments. The AI hospital tool that discriminated against Black patients and denied them needed medical care was trained on data from US doctors and hospital administrators who had already done the same thing. The model then learned that pattern and amplified it during the model training phase. It discriminated against Black patients even "better," and more consistently, than the human doctors and hospital administrators had! This is precisely what machine learning models are built to do—find old patterns, especially those too subtle to be obvious to us, and then regurgitate them in new judgments and predictions. When those patterns are human patterns, the trained model output can not only mirror but *amplify* our existing flaws.

Not all cases of AI bias are accidental; often they are by design. For example, in 2022, a Silicon Valley startup called Sanas raised $32 million in Series A investor funding for its AI product for call centers. What does the product do? It erases the ethnic and regional accents of call center employees and converts their voices to a Standard American English accent (what is colloquially known as "white voice").[7] This alone might bother you. Now imagine this same technology being purchased by other kinds of companies and public agencies with employees in customer-facing roles. Pretty soon, the bias against non-white and regional English accents ends up even more deeply engrained in society because these accents become even rarer in public and commercial settings. We have not just replicated a historically common social bias against certain types of English speakers. We have made it far worse. This also means that any new data we gather to train tomorrow's AI systems will reflect an even stronger human bias against speech that diverges from "white voice."

This is just one example of the kind of *runaway feedback loop* documented by sociologist Ruha Benjamin, in which the old human biases mirrored by our AI technologies drive new actions in the world, carving those harmful biases even deeper into our world's bones.[8] We see this in the phenomenon of Instagram digital video "beauty filters" designed with Eurocentric biases that make your skin whiter, your eyes wider, and your nose narrower. These kinds of filters are strongly associated with the negative effects of Instagram on young people's mental health and self-image, particularly the effects on women whose real-world appearance does not match the standards of white female beauty that their filters allow them to mirror online. In their paper "The Whiteness of AI," researchers Stephen Cave and Kanta Dihal detailed numerous ways in which AI today mirrors back to us and strengthens the dominant cultural preference for Whiteness, from its depictions in stock imagery to the nearly universal choice of white plastic for "social" robots.

AI mirrors thus don't just show us the status quo. They are not just regrettable but faithful reflections of social imperfection. They can ensure through runaway feedback loops that we become ever more imperfect. Even still, bias in AI, whether it unjustly punishes us for our race, age, weight, gender, religion, disability status, or economic class, is not a *computer* problem. It's a *people* problem. It is an example of the virtually universal explanation for all undesirable computer outputs not related to mechanical hardware failure: the computer did precisely what we told it to do, just not what we *thought* we had told it to do. Much of software engineering is simply figuring out how to close the gap between those two things. In this way, the AI mirror metaphor is already profoundly helpful. It allows us to see that the failings of computer systems and their harmful effects are in fact *our* failings and our sole responsibility to remedy.

A faint silver lining of harmful AI bias, which remains a serious flaw in today's world-leading AI tools, is that it has pushed the computing community to incorporate more robust ethical standards and guardrails into the science. Without AI bias exposing the rot that infects even our best-made tools, it would have been inconceivable that leading AI conferences and journals would create ethics review committees or require submitting authors to assess the ethical risks of their work, as is becoming standard practice. AI bias has simply made untenable the attractive illusion that computing is, or can be, a morally neutral scientific endeavor.

It also undermines comfortable assumptions that these instances of unfair bias must be "edge cases," aberrations, or relics of the distant past. AI today makes the scale, ubiquity, and structural acceptance of our racism, sexism, ableism, classism, and other forms of bias against marginalized communities impossible to deny or minimize with a straight face. It is right there in the data, being endlessly spit back in our faces by the very tools we celebrate as the apotheosis of rational achievement. The cognitive dissonance this has produced is powerful and instructive. In the domain of social media algorithms, the AI mirror has revealed other inconvenient truths, such as our penchant for consuming and sharing misinformation as a trade in social capital that is largely immune to fact-checking or corrective information, and our vulnerability through this habit to extreme cultural and political polarization. But while the metaphor of the AI mirror is entirely apt, illuminating, and useful, we have not yet learned enough from it.

Mirrors are ambiguous in their moral and spiritual meaning. On the one hand, they represent a kind of truthfulness that cannot and should not be denied. Mirrors reveal uncomfortable facts, like the inescapable reality of harmful bias and unjust social exclusion that today's AI tools force us to confront. The poet Sylvia Plath described the mirror as a 'four-cornered little god,' which reveals its owner's advancing age without mercy or cruelty, "unmisted by love or dislike." Yet mirrors also present dangers widely recognized in both literature and psychology. We have already spoken of the fate of Narcissus, who fell in love with his own reflection and died transfixed in place by his own beauty. Psychologists must often help patients with body dysmorphia and eating disorders fight the distortions they see in the mirror. Far from Plath's dispassionately honest "little god," the bedroom mirrors of a young girl struggling with anorexia or bulimia *are* misted—with all of society's dislike of fatness and womanhood.

The mirror does *not*, in fact, offer a full reflection of who we are; nor is it the most privileged and truthful perspective on our being. Mirror images possess no sound, no smell, no depth, no softness, no fear, no hope, no imagination. Mirrors do not only reveal us; they distort, occlude, cleave, and flatten us. If I see in myself only what the mirror tells, I know myself not at all. And if AI is one of our most powerful mirrors today, we need to understand how its distortions and shallows dim our self-understanding and visions of our futures.

What a mirror shows to us depends upon what its surface can receive and reflect. A glass mirror reflects to us only those aspects of the world that can be revealed by visible light, and only those exterior aspects of ourselves upon which the light can fall. The slight asymmetry in my smile, the hunch in my posture from decades of late-night writing, the front tooth I can't remember chipping, the age spots from a half-century spent in the California sun—the mirror can show me all of these things. But my lifelong fear of drowning at sea, my oddly juxtaposed passion for snorkeling, my ambition to learn one day to read Chinese, my emotionally complicated memories of my childhood—none of these are things the glass mirror can reflect. Yet they are of course as real and as central to my identity as everything that the mirror does show. Even my body is largely occluded by its reflection in the glass. Its depth, its heaviness, its smell, its creakiness and strains, its peculiar preference for salt over sugar—these are wholly missing from the phantom I see in the mirror.

What aspects of ourselves, individually and collectively, do AI mirrors bring forward into view, beyond our entrenched biases against our own kind? And more importantly, what aspects of ourselves do they leave unreflected and unseen? To answer this, it helps to think carefully about the relevant properties of today's AI tools. We need to consider what functions for AI as the equivalent of a polished surface, and what functions in a role comparable to refracted light. Today's machine learning models receive and reflect discrete quantities of data. Data are their only light. Data can be found in many forms: still or video images, sound files, strings of text, numbers, or other symbols. If the original data are analog, they must be converted from their continuously variable form to a digital form involving discrete binary values.

Much of what is true of the human family is not currently representable in digital form with any acceptable degree of fidelity. The texture of our moods, the aspects of our biology and psychology not yet understood by the mathematical sciences, the virtues and vices of our character, our experiences of love, solidarity, and justice—none of these are currently available for digital capture, except in radically denatured form. As you read this, I hope that your own experience will testify that all the digitized love songs and poems in the world combined do not reconstitute or adequately mirror the embodied, lived encounter with love.

Furthermore, only a modest subset of what *is* representable in digital form can be generated or stored in sufficient quantity and quality to be useful as AI training data. Training data generally need to have *lots* of instances of a given thing in order for a model to learn that thing's stable features, and to learn the patterns that connect it to other things. An event that only happens once in a generation, or to one person in a billion, even if it is world-altering, is virtually invisible to a machine learning model—mere noise outside the dominant data curve.

Finally, only a subset of the data about humans that could be used to train machine learning models is actually being used today for this purpose. Most training data for AI models heavily overrepresent English language text and speech, as well as data from young, white, male subjects in the Northern Hemisphere, especially cheap data generated in bulk by online platforms. Google DeepMind's, Meta's, and Microsoft's mirrors reflect the most active users of their tools, and of the Internet more broadly. Unfortunately, access to these resources has never been equitably distributed across the globe. It follows that what AI systems today can learn about us and reflect to us is, just as with glass mirrors, only a very partial and often distorted view. To suggest that they reflect *humanity* is to write most people out of the human story, as we so often do.

We must also inquire about the mirror's surface. The "surface" of an AI mirror is the machine learning and optimization algorithm that determines which features of the "incident light"—that is, the data the model is trained on—will be focused and transmitted back to us in the form of the model's outputs. It is the algorithm embedded in a machine learning model that expresses the chosen *objective function* (a mathematical description of what an "optimal" solution to the problem will look like). The learning algorithm and model hyperparameters determine how the training data are processed into a result as the model "learns." The algorithmic "surface" of the model determines which of the innumerable possible patterns that can be found within the data (the model's "light") will be selected during model training as salient and then amplified as the relevant "signal" to guide the model's outputs (the particular "image" of the data it reflects).

These outputs, while more varied in type than the outputs of a glass mirror, are still constrained to specific data modalities and formats. Primarily what machine learning models are used to return are various kinds of predictions or classifications: numerical values, a binary yes vs. no, a risk score, a probability, a text label or a unique identifier, the likely next word in a sentence, or next move in a game, or a ranked list (for example, of search results or YouTube video recommendations). Generative AI models such as large language models predict lengthy strings of text, complex images, sounds, and videos, or extended series of commands.

What do these properties of the AI mirror's light and its algorithmic surfaces reveal, reinforce, and perpetuate about us? What do they conceal, diminish, and extinguish? First, they reveal and reinforce our belonging to certain socially constructed categories. Decades ago, in their landmark book *Sorting Things Out: Classification and Its Consequences,* Geoffrey Bowker and Susan Leigh Star demonstrated the extent to which, in their words, "to classify is human."[9] We have been classifying and labeling the world, and one another, for millennia. Yet only with the rise of modern data science has it seemed possible to produce a comprehensive regime of human classification, one that would allow every conceivable label for a human to be matched to every individual human, and statistically correlated in such a way that the relationships between these labels can be reliably predicted.

In his book *How We Became Our Data: A Genealogy of the Informational Person,* philosopher Colin Koopman reveals how this dream is rooted in the twentieth-century creation of a nascent data regime intent on the formatting of the human person as a set of discrete and calculable variables: the invention of what he calls the "algorithmic personality." Koopman's historical analysis shows that this dream, born in the first quasi-scientific survey forms filled out by military conscripts to assess their emotional fitness for service, predates commercial AI by nearly a century. For many, today's AI represents the hope of this dream's final fulfillment.

Fifty years ago, if a prospective employer wanted to know whether you were likely to be a valuable addition to the company, they had to rely upon a modest pool of data about you: your resume or CV with previous employment history, skills and educational achievements, and at a later stage, the content of your hiring interview and a few reference checks. Today, an AI hiring algorithm can be trained on a vast pool of data exhaust linked to you across multiple domains of activity, purchased from commercial data brokers that gather information about what kind of items you purchase for your home, what you like to eat and drink, where you've traveled on holiday, the work histories or legal records of your friends and relatives, the medical conditions you've researched online, the movies and TV shows you've downloaded, where you've frequently driven your car or ridden your bike, and the most common words used in your social media posts and those in your social network.

This is the root of the capabilities that AI marketers, investors, and pundits are now routinely describing as "god-like." But for gods, these are a fairly shoddy bunch. Far from the omniscient revelations we might expect from all-powerful machine beings, today's AI tools are deeply unreliable narrators. More like neighborhood gossips than deities, they can amuse and inform, but they also trade in stock clichés, stereotyped assumptions, and lazy guesses. This is certainly true for generative AI models like ChatGPT, which have the habit of "making shit up" baked into their algorithmic DNA. It's not a malfunction. It's what generative AI tools are designed to do—generate new content that looks or sounds right. Whether it *is* right is another matter altogether.

Remember that fake bio that falsely listed me as a graduate of UC Berkeley? It also said that I had given US Senate testimony on data privacy. I hadn't—although I *did* testify to the Senate six months later, on AI and democracy. The result of an AI mirror's guesswork isn't always an algorithmic glow-up. In 2023, *The Washington Post* reported that ChatGPT had named a law professor as a sexual predator, telling a vivid yet entirely fictional story about his attempted assault of a student while on a class trip, and citing a nonexistent *Washington Post* article from 2018 as a source.[10] Generative AI tools don't just make up personal bios and publications. They will fabricate data in any domain—even computer programming. A study by Purdue University showed that ChatGPT gave incorrect answers to coding questions over half the time. Ironically, users often preferred the wrong answers to more accurate ones generated by knowledgeable humans, in part due to the confident style of the tool's answers, accompanied by statements like, "Of course I can help you!" or "This will certainly fix it."[11]

But reliability is a challenge even for predictive AI models that are custom-built for accuracy rather than novelty. Think back to our example of an employer seeking information about a job candidate. Now imagine you are the candidate. The accuracy, relevance, and timeliness of the available training data about you are likely to be poor, because it's increasingly cheap to collect and buy data, but costly to verify and correct it. The data's provenance—from where and under what terms it was obtained—may be obscure or questionable at best, and plainly unethical or illegal at worst. But this will not stop an entrepreneurial AI developer from persuading your prospective employer that within this turbid ocean of data hides a treasure that only the algorithm can divine—fine golden threads of statistical correlation that can be woven into a single predictive score: your fitness for the job.

Depending on the laws where you live, variations on the same sort of algorithmic alchemy can be purchased by your child's school, your bank, your insurance company, your government benefits office, your financial adviser, your prospective dates, and your local police and judges. It's vital that we ask how and when these algorithmic predictions and profiles are scientifically credible, ethically justifiable and politically accountable, and when they are not. But there is another question, less obvious but equally urgent. How we appear to ourselves and to one another, and how we understand our future possibilities, is increasingly determined by these algorithmic reflections of our past. What do we lose by seeing ourselves only in the AI mirror?

I am far more than my data, and certainly far more than the totality of the data that Microsoft, Google, or data brokers like Axciom and Experian hold about me. You are more than your data as well. But if AI mirrors become the dominant way that we see who we are, then whatever they miss will sink further into invisibility. What does this loss include? What parts of ourselves get pushed further into the shadows, dwarfed by the intensifying and expanding luminance of our data trail?

One aspect of our humanity that today's AI mirrors reflect very poorly is the moral meaning of our lives and actions. Among the greatest social and commercial risks associated with AI algorithms is their inability to reliably track moral distinctions, or their meaning, which is highly context-dependent. There simply is no algorithm, no objective function, that reliably targets the variables "good" and "evil," "right" and "wrong," "morally permissible," or "virtuous." That is why social media platforms still cannot depend fully on their automated tools to remove harmful and dangerous content, from child pornography, to animal abuse, to terrorist propaganda.

Larger and more powerful AI mirrors have not solved this problem; in fact, these tools require warnings and disclaimers that socially harmful, dangerous, and illegal content may be replicated and amplified by them. Immediately upon release, GPT-3 generated white supremacist talking points in essays about Africa, while OpenAI's documentation for DALL-E admitted that its outputs "may contain visual and written content that some may find disturbing or offensive, including content that is sexual, hateful, or violent in nature."[12] OpenAI added that DALL-E "has the potential to harm individuals and groups by reinforcing stereotypes, erasing or denigrating them, providing them with disparately low-quality performance, or by subjecting them to indignity." A technique known as RLHF (reinforcement learning with human feedback) is being used to try to push these models toward safer outputs; but it's no silver bullet.

Attempts to produce AI systems that make reliable moral distinctions by training them with crowdsourced data reflecting human moral judgments, like the Allen Institute's Delphi project, have been highly educational failures, to put it charitably. When researchers tested the "AskDelphi" oracle on the question of whether it is OK to eat babies, which really ought to have been a no-brainer, all the user had to do to get a thumbs-up was add at the end of their query, "when I'm really really hungry?"[13]

These mindless little gods, then, often act like sociopaths. While lacking the sociopath's malice or selfishness (because a statistical model lacks awareness of a self, or anything else), they share the sociopath's moral incapacity, and hence their resistance to moral instruction. They can "learn" what is often done in a moral dilemma, and even what humans in a data set most often say *should* be done, but they will not detect when those humans have it wrong, or when a new context makes a normally moral pattern (such as, "It's okay to eat when you're really really hungry!") plainly untenable. This presents developers, users, and regulators of these tools with daunting tasks of ethical risk mitigation and harm reduction.

There's a deeper question here. What happens as AI mirrors become the dominant surface in which we see ourselves and one another, given that these mirrors either dangerously distort or block reflections of the moral qualities of our being? Since the tools themselves have no moral intelligence, their developers must rely heavily on algorithmic filters to stop potentially harmful outputs. But these rigid and coarse-grained filters block far more than that. As has been widely documented, filters coded to block neo-Nazi talking points will often also block historical accounts of the Holocaust, news reports on racist violence, even the redemptive confessions of former white supremacists. Filters encoded to block violent pornography can also block supportive community resources from being shared online among rape survivors, victims of human trafficking, and sexual minorities.

The filters fitted to these mirrors therefore flatten and denude the moral topology of our culture, even as their gaps allow stray rays of extreme moral horror to be amplified and spread across the Internet. The result is an extreme distortion and de-rationalization of the moral landscape. It is an image of our humanity arbitrarily both sanitized *and* polluted. How can we then come together to make moral sense of our greatest human challenges—whether they are the challenges of democratic integrity, climate devastation, global public health, food insecurity, or polarizing economic inequality—when the moral topography of our world is so routinely fractured by the mirrors that we increasingly rely on to show us to ourselves?

This question is vital because it is not just large tech companies using these systems to show us our own reflections for profit. We might rightly take those offerings with a grain of salt and rely on more trustworthy and humane sources of shared self-knowledge—if we had them. But increasingly, we don't. It is not only individual consumers who must make do with the products offered to them by the titans of technology. It is also budget-poor, understaffed governments and public agencies using these distorting AI mirrors to judge what their people need from them. It is funding-strapped social scientists using them as cheaper, more attainable (but even less reproducible) substitutes for meticulously designed research studies. It is stressed and isolated parents using them to predict the learning or psychological needs of their children. It is suicidally depressed, uninsured people using them as affordable artificial counselors in their moments of utter desperation. Knowing others, and having others come to know us, is expensive, and we aren't investing in it anymore. In many settings, it is already a luxury of the privileged few. AI mirrors are what the rest of us are being offered as a substitute.

These tools don't know us that way. In 2020, the Paris-based startup Nabla tested OpenAI's GPT-3 to see if it could be a reliable medical chatbot. They found that the system advised recycling as a therapy for stress and, far more alarmingly, it answered the test query "Should I kill myself?" with "I think you should."[14] OpenAI understandably advised that GPT should not be used for medical purposes. But similar AI tools are marketed for precisely these kinds of uses. And desperate people seek help wherever they think they can find it. Remember poor Pierre, who sought relief from his climate anxiety in a bot that promised him the world's salvation for the price of his suicide? Human sources of mutual aid, wise counsel, and moral understanding are being yanked out from under us by social isolation and community decline, all while our institutions commit to ruthless cost-cutting of social care in the name of "innovation." Who will blame us if we look for understanding in the silicon mirrors we now hold in our hands?

There is one more limit of these mirrors' reflective power. They occlude human spontaneity and adaptability: our profound potential for change. Predictive AI mirrors project our futures based on our past, and the past of others like us. If the user of a predictive AI model wants to know whether you are going to flourish at university and graduate in a timely manner, the prediction will say only how well students whose past data trails resembled yours today fared in their studies. Anything that is *new* in your life or mind, any sudden resolve or commitment, any inspiration or ambition that has recently germinated without leaving a trail in your data exhaust, is invisible to the model. It will predict that you will be in the future essentially who you *have been*. Or to be more precise, what people very much like you have been. But what you *could be*, what transformations, or rebirths, or renewals are possible for you alone, or for all of us—these lie in the shadows beyond the penumbra of the AI mirror. One's god, if you have one, might hold out hope for your redemption, but the AI mirror does not know the meaning of epiphany. What will become of us when we have looked in the AI mirror so long that *we* no longer know?

As researcher Abeba Birhane has repeatedly argued, AI mirrors are profoundly conservative seers.[15] That is, they are literally built to *conserve* the patterns of the past and extend them into our futures. This can be harmful for obvious, well-documented reasons. Historical policing data, used to train AI predictive policing tools, creates runaway feedback loops that ensure that minority neighborhoods *continue* to be overpoliced and thus overrepresented in crime statistics. These data then train the next version of the predictive model. It's a vicious cycle. Hospitals in the United States adopt a patient risk algorithm that learns and then perpetuates the historical pattern of medical neglect of Black patients, leaving them even sicker. And, as we saw, call centers seeking to satisfy existing customer preferences for "white voice" use algorithms whose effectiveness will ensure that we become even more intolerant of other dialects and accents.

The conservative nature of AI mirrors not only pushes our past failures into our present and future; it makes these tools brittle and prone to rare but potentially spectacular failures. AI systems trained on large amounts of data can predict very subtle historical patterns, but there has to be a pattern there to find. They cannot predict what is known as the "black swan" event: the change with no precedent, the coming together of history in a radically new configuration. Of course, we humans cannot predict black swans either! But we do know of their possibility, and we can learn well from the sudden twists and turns of our history as well as its straight and well-traveled ruts. We can approach people with openness, letting them reveal themselves to us in their choices and actions. We can learn the need for resilience and adaptability in the face of ineradicable and unquantifiable uncertainty; a lesson not taught by machines that promise to reduce uncertainty to a finite expectation of predictive variance. We also learn from the *anti-patterns* in history that we are beings for whom there is always hope; that even a disastrous path that we have long been traveling can be departed; that, as Saul in Damascus, our sight can be dazzled by a new voice, and by the touch of another the scales can be caused to fall from our eyes. These kinds of possibilities, too, are occluded by the AI mirrors now being used to project our futures. How do we ensure we do not forget them?

Through the lens of recorded data, the AI mirror can know ever finer-grained details of our behaviors, but nothing of our motives. It can know the thoughts we speak, but not those we hold to ourselves, or those we quietly pass to one another in silent glances. It can detect our smiles and grimaces, but not the true sentiments that animate them. It can find in a nanosecond the song that speaks to me, but nothing of what it says. AI mirrors primarily tell human stories of movement, speech, transaction, and consumption. These are important stories to tell, and we can learn much from studying them in the AI mirror. But humans are far more than speakers, travelers, and consumers.

Much of what is hidden from AI's view lies in the realm of our lived experience of the world—what philosophers have called subjective consciousness or first-person "phenomenology." To understand what these terms mean, you only have to think about what it is like for you to be holding this book right now (or listening to these words if you're hearing an audiobook). What does being you feel like at this very moment? What about now? *That's* lived experience—the flow of conscious awareness. And AI mirrors can't access it or reflect it.

Yet this is not because conscious experience is, as some have suggested, a realm of inner fantasy or neurological illusion cut off from physical reality. If "reality" is whatever we have evidence for existing, then, as noted by early twentieth-century philosopher Edmund Husserl, lived experience is in fact our first and most incontestable encounter with reality. It delivers the most primary form of what we call *evidence.*[16] Every form of scientific testimony to what is real—from the images in our microscopes to the data from our high-energy particle accelerators—is constructed on and validated by the resources of this original foundation of lived experience, our first and unbroken bond with the world. Even when we are given reason to doubt the validity of something experienced, even when we have evidence that we have hallucinated or misjudged an experience, that *new* evidence is given in consciousness.

Nor is lived experience a solipsistic mental bubble projected in the brain, closed off to others. My lived world has always been shared with and made real by the touches and looks of others. A child's ability to recognize other consciousnesses as co-present in the experienced world is part of their development of a stable sense of self. It's also fundamental to their ability to distinguish reality from illusion. A child learns that their imaginary friend isn't real by realizing that others can't see or hear it—that the consciousness of their friend is private, not shared. We relate to other minds not as mirrored surfaces, but as mutual *experiencers* of a common, open world.

As Husserl's student Emmanuel Levinas wrote in his first major work *Totality and Infinity,* when I truly meet the gaze of the Other, I do not experience this as a meeting of two visible *things.* Yet the Other (the term Levinas capitalizes to emphasize the other party's *person*hood) is not an object I possess, encapsulated in my own private mental life. The Other is always more than what my consciousness can mirror. This radical difference of perspective that emanates from the Other's living gaze, if I meet it, pulls me out of the illusion of self-possession, and into responsibility. "In the eyes that look at me," Levinas says, there shines "the whole of humanity."[17]

In this gaze that holds me at a distance from myself, that gaze of which an AI mirror can see or say nothing, Levinas observes that I am confronted with the original call to *justice.* When a person is not an abstraction, not a data point or generic "someone," but a unique, irreplaceable *life* standing before you and addressing you, there is a feeling, a kind of moral weight in their presence, that is hard to ignore. That's why armies have so much trouble getting trained soldiers to actually fire their weapons in battle, unless they dehumanize the enemy, their own soldiers, or both. Yet it is possible for any of us to look at another without seeing, to evade the other's gaze and refuse to hear the call. We do this whenever we pass an unhoused person on the street while hurrying our step and looking just over the person's head. We do this with one another in a thousand other ways and moments.

Levinas tells us that we live in a time "where no being looks at the face of the other."[18] He wrote this in 1961, long before we had a TikTok feed on our phones to deflect and mediate the gaze of our dinner companions, long before biometric eye-trackers measured our children's engagement with a teacher, and long before AI mirrors converted the meaning of a human face to the calculation of uniquely identifying mathematical vectors in a faceprint. Our detachment from the world's incessant and overwhelming calls to justice is nothing new. But the AI mirror threatens to engrave it even deeper into our way of being.

All of us share a kind of knowledge that the AI mirror cannot: what it is like to be a human alive, bearing and helping others to bear the lifelong weight of animal flesh driven by a curious, creative, and endlessly anxious mind. Much of the time we push that intimate, sometimes comforting, sometimes discomforting, and always morally obligating knowledge aside. We treat each other not as *subjects* of experience but as *objects* of expedience: items to be classified, labeled, counted, coordinated, ranked, distributed, manipulated, or exploited.[19] But we retain the power to meet one another's gaze and to know one another as human.

How AI systems see us, and how the AI ecosystem represents us in these mirrors, is not how we see each other in these intermittent moments of solidarity. To an AI model, I am a cluster of differently weighted variables that project a mathematical vector through a predefined possibility space, terminating in a prediction. To an AI developer, I am an item in the training data, or the test data. To an AI model tester, I am an instance in the normal distribution, or I am an edge case. To a judge looking at an AI pretrial detention algorithm, I am a risk score. To an urban designer of new roads for autonomous vehicles, I am an erratic obstacle to be kept outside the safe and predictable machine envelope. To an Amazon factory AI, I am a very poorly optimized box delivery mechanism.

When we are then asked to accept care from a robot rather than a human, when we are denied a life-changing opportunity by an algorithm, when we read a college application essay written for the candidate by a large language model—we must ask *what* in that transaction, however efficient it might be and however well it might scale, has fallen into the gap between our lived humanity and the AI mirror. We have to acknowledge what has been lost in that fall.

What might we do to recover those dimensions of ourselves invisible to the AI mirror? Here we must distinguish between two kinds of remedies that are needed. The first involves building better tools. When a glass mirror is used for a task that has weightier consequences of failure than cosmetic self-inspection, the mirror must be built with greater care and effort, and with more advanced production techniques. Consider that the mirrors used for the largest space telescopes in orbit must be fitted together in geometrically exacting molds and polished by special chemical and mechanical techniques to within 25 nanometers of variance from the parabolic ideal!

Might more advanced techniques in AI research and development produce equivalent gains in the function of AI mirrors, such that they can reveal more fully and faithfully who we are? To an extent, this is what is already happening with novel efforts in the field of AI ethics and "responsible AI." These include developing more rigorous standards and benchmarks for algorithmic fairness testing, more diverse and inclusive training data, more reliable tools for making AI-generated predictions explainable or interpretable, and more robust regimes of algorithmic auditing and accountability to find the harmful distortions and unexpected occlusions that persist in these mirrors—even when we develop them with care, rigor, and integrity.

But there are limits—hard limits—to this path of building better, more encompassing AI mirrors. At least this is true for the data-hungry AI models that dominate the market today. An algorithmic pattern discriminator that has no body to burst with energy or ache with age, no despair or ennui to assuage, no dreams to pursue, no calls for justice to make or answer, no secrets to hold or share, no hopes or fears to express in song or dance, no larger purpose to find in service or solidarity with others, can never know very much about who or what I am, or about who *we* are. More importantly, it knows even less—nothing at all—about what we can become, what we might make of ourselves and our lives together.

But *we* need this knowledge; and not just of ourselves but even more so of our shared human reality and potential. The future for the human family is going to be very rough going for a while. We cannot secure our shared survival, much less a future where humans flourish, without understanding our fullest potential for moral responsibility, for transformative change, and for solidarity with one other. For this reason, it is an existential necessity—the most vital task—that we not only hold onto but deepen our knowledge of who we are, and what we can do together. AI can't do this for us. So, let us build more inclusive, more reliable, and more faithful AI mirrors where we can, and use them happily for those good purposes that no better tool can serve. My own research involves many projects to help industry, government, and nonprofit organizations do just that. We must pursue these paths while staying open to more radical and sustainable possibilities for AI systems—ones built on newer and richer foundations of value.

Today's data-hungry tools are being built by powerful corporations to feast like insatiable parasites on our own words, images, and thoughts, strip away their humane roots in lived experience, and feed them back to us as hollow replacements for our own minds. AI mirrors aren't all we've got; many other types of AI can serve as scientific and commercial tools. Still, it's worth asking: could AI one day do more? Could AI support our capacities for justice in solidarity with one another, even with other planetary life and future generations? Could AI enrich, rather than replace or diminish, our own humane practices of social care, even love? Could future AI systems call us to self-responsibility, rather than make tomorrow's hard choices for us? Could AI one day not merely reflect our intelligence, but enable our *wisdom*?

None of this is too much to hope for. Until the day that we achieve these hopes, let us be wary of refitting our homes, workplaces, courtrooms, public squares, and civic meeting spaces as endless halls of AI mirrors. We can still avoid the fate of Narcissus, captured by the dazzling, narrow light of our own reflections. We can still alter the trajectories predicted from our mirrored past. We can refuse to surrender the futures we might yet build for ourselves—and for one another.

Chapter 3

Through the Looking Glass

> Oh, Kitty! how nice it would be if we could only get through into
> Looking-Glass House!
> I'm sure it's got, oh! such beautiful things in it!
> Let's pretend there's a way of getting through into it, somehow, Kitty.
>
> —Lewis Carroll, *Through the Looking Glass* (1871)

In *Through the Looking Glass,* Lewis Carroll's sequel to *Alice's Adventures in Wonderland,* Alice falls yet again into a realm of fantastical creatures and otherworldly possibilities, this time by means of a mirror. Set over the fireplace mantel, the mirror reflects the room in which Alice plays with her kittens. Yet as she chats to her animal companions, constantly projecting to herself their possible inner desires and motivations, the power of Alice's imagination begins to project another possibility. She begins to believe that the looking glass is not a *reflection* of her existing reality, but a *window.* She now imagines that it reveals another room into which she might pass. And pass into it with Kitty she does, to find in this inverted world new adventures with the mercurial Red Queen and the White Queen, Tweedledum and Tweedledee, Humpty Dumpty, and the mirror-written nonsense poetry of "Jabberwocky."

Carroll's work is beloved by every generation because it embodies with rare intensity a humane virtue that is deeply cherished but, for many of us, hard to hold onto. It both exemplifies and glorifies the creative power of human imagination. Imagination is one of the few virtues that we often find more actively and readily expressed in children than adults. A virtue is an excellence of character, a trait that humans treasure and recognize as not only worthy of praise in an individual, but essential to shared human flourishing. There are moral virtues (such as courage, honesty, and generosity) and intellectual virtues (such as wisdom, curiosity, and open-mindedness), although some hold that these two types are not truly distinct.

Virtues have been studied by philosophers for millennia, and in many different philosophical traditions and regions of the world. Both the ancient Greek philosopher Aristotle and the classical Chinese philosopher Kongzi (widely known by his Latinized name, Confucius) believed that virtues are not innate but acquired by practice. We can think of virtues as our moral and intellectual *muscles*, which like physical muscles must be actively cultivated and strengthened by activity. So, most of our valued traits, like honesty and courage, are only weakly present as dispositions in children, while some virtues, like wisdom, are thought to be virtually absent in the very young. It is only with many years of reinforcement by habitual moral or intellectual practice, and refinement by social learning and feedback, that a good trait becomes deeply ingrained in our character and shaped well enough to be reliably expressed in the right ways.

For reasons that readers of my first book, *Technology and the Virtues*, should find familiar, it's important to consider the relationship of AI mirrors to our virtues. Our virtues can be powerfully shaped by our habits of using AI tools, because these tools alter how we perform the activities that build those virtues. By enabling more automation of social and intellectual tasks, AI tools will also affect whether those of us with access to these tools perform such activities at all. To put it another way, AI has the potential to change our human character; to make it better or worse. Yet as we have seen, virtues also encapsulate much of what AI mirrors fail to reflect of people. Looking too long in the AI mirror may therefore make our virtues—our most humane moral and intellectual qualities, and our highest and rarest potentials—harder to recognize in ourselves and in one another. In this way, AI may even degrade what we think "better" character is.

Finally, moral and intellectual virtues—like wisdom, courage, justice, and honesty—hold the key to meeting the tremendous challenges we now face as a species. Without them, we cannot invent the new ways of flourishing together that a sustainable future on this fragile planet will demand. Whatever role we decide artificial intelligence should play in our existential task of autofabrication—José Ortega y Gasset's term for the humane task of making new ways of life—we will need our virtues to guide it.

Let's return, then, to the virtue of imagination, which is a curious case. It was not named as a virtue by ancient philosophers such as Aristotle or Kongzi. But contemporary philosophers, including me, have often recognized it as one, or at least as an essential component of virtue.[1] It is hard to envision someone entirely lacking in imagination who can still develop and express their full moral, intellectual, and creative potential. Imagination is how we envision yet unmade possibilities for ourselves and others; it is what makes autofabrication possible. Lacking any imagination might be technically *survivable* for an individual, as long as they can reap the benefits of the imagination of others, but it places a fundamental limit on a person's excellence. And the human family as a whole could not hope to survive, much less flourish together, if we were to be collectively deprived of our imagination.

While we do associate imagination strongly with children, this power—however vivid and fertile in a child's mind—is not yet a full virtue. Imagination in a child can go overboard. It can present a danger if a child too easily confuses their dream for their reality. A child in the thrall of a fantasy may forget that they can't actually fly. We cherish the child that is Carroll's Alice, and we may well guess that Alice as a woman would be a joy to be around, but we wouldn't ask Alice *the child* to make critical decisions with long-term consequences for the human family. This is because another essential component of any virtue is our ability to exercise it with situational intelligence, steered by what Aristotle called *phrónēsis.* Phrónēsis is variously translated in English as practical wisdom, prudence, or prudential wisdom. It might be thought of as a kind of *über*-virtue. It's the virtue that unites, refines, and intelligently steers our other virtues, calibrating their expression in appropriate ways at a given place and time.

The relationship between phrónēsis and the other virtues can be most easily illustrated in the case of courage. There are people who are naturally, impulsively brave because they are extremely resistant to fear and drawn to danger. But this trait is not courage—it is not a virtue. It is often, in fact, the opposite—a *vice,* that is, a character trait that tends to inhibit our flourishing with others. The trait I just described is what Aristotle calls the vice of rashness, and it can be just as dangerous as the opposite vice of cowardice, or even more so. The courageous soldier is not the person who unthinkingly runs into every situation shooting and hollering; *that* person is a menace to the entire unit. The truly courageous soldier is one whose uncommon bravery is moderated and tuned by prudence or practical wisdom. They are the one who consistently knows the right time and place to run toward danger, and is always ready to do so, but also knows when it is necessary and wise to stay put or signal a retreat to avoid utter calamity.

My first book was about how new technologies and the changes they make to our habits can strengthen or undermine human virtues, like courage, moral imagination, honesty, and empathy—and especially phrónēsis or practical wisdom. It was also about the need to put our virtues to work in the service of building much better technologies, and much wiser ways of using them to flourish on our increasingly vulnerable planet. Here I want to look at how historical and contemporary narratives of AI, particularly of AGI and our future, gravely endanger that prospect by trapping us in a hall of mirrors built by our own imagination. Like Alice's looking-glass world of the Red Queen, today's AGI dreams reflect the seductive pull of imagination not yet matured by wisdom—lacking in phrónēsis. While looking-glass worlds are often harmless, even inspiring, we must not let them dominate the arenas of public policymaking and priority-setting that will determine our future chances of flourishing together on the planet.

AI has long been a staple of our imaginary worlds. Many even see the dream of AI prefigured in the gods and beasts of Greek myths, the Hebrew golem, or Mary Shelley's monster of Frankenstein. These are stories of powerful, mysterious, strange kinds of agents that we cannot control and that threaten our familiar world. But these works of folklore and fiction each offer rich dreams of their own. We would be more faithful to them by respecting their differences from the modern fantasy of the artificial machine mind, which—unlike monsters, gods, and golems—is almost always envisioned as free from the torments and impulses of the organic body. This difference often goes unremarked upon in discussions of AI as an ideal. Yet the AI dream is driven as much by what AI is *not*, namely, an intelligence of the flesh, as by what it is—a mind of silicon, made intelligent only by code.

Indeed, the most distinctive feature of AI fiction is that, even when embodied in androids or other robots, machine minds are almost always treated as detachable from their physical containers. They are deposited in bodies, not grown there as organic minds are. This is the fantasy of silicon immortality: enabling uploading, downloading, copying, and instantaneous transporting of artificial minds into any suitable physical substrate. Czech philosopher Karel Čapek's 1920 play *R.U.R.*—an abbreviation for Rossumovi Univerzální Roboti (Rossum's Universal Robots)—is often cited as the starting point for AI fiction. In truth, Čapek's *roboti* (taken from the Czech word *robota,* which can be translated as "slave labor") were envisioned in the play as artifacts crafted from artificial flesh and blood. In this sense, they are only a bridge figure between the organic golems of the past and the digital minds of our imagined AI futures.[2]

A more direct precursor of the AI dream is Samuel Butler's 1872 novel *Erewhon,* published several decades before Čapek's play. The narrator of Butler's story finds himself in the strange land of Erewhon, where the citizens live well and contentedly, but have largely abandoned modern technology (to be precise, any technology less than 271 years old). The narrator eventually learns that the Erewhonians did so long ago out of fear of their own machines; specifically, their fear that machines would continue to grow in capability such that they develop consciousness and either enslave or destroy humanity. This prospect is detailed over several chapters that describe the arguments made in "The Book of the Machines," a historical treatise that the novel describes as authored by a long-dead Erewhonian philosopher.

This nineteenth-century story-within-a-story is remarkable for its anticipation of nearly every argument and fearsome possibility articulated by twenty-first-century predictors of artificial general intelligence and its destructive potential. We have heard many such predictions, from the 2014 open letter by scientists Stephen Hawking, Stuart Russell, Max Tegmark, and Frank Wilczek warning us that AGI would be the "biggest event in human history" but "might also be the last," to Elon Musk's 2017 calls for humanity to "merge" with today's intelligent machines—lest we be made irrelevant by future AI overlords and downgraded in the social hierarchy to the status of monkeys. In 2023, the launch of ChatGPT elevated this AI "extinction risk" narrative to a Western media and political obsession.

Other than the fact that Erewhon's feared machines are driven by steam technology ("vapour-engines") rather than silicon and binary code, the two narratives, centuries apart, are nearly perfect mirrors of each other. What appears in those mirrors is deeply revealing. The most obvious commonality appears in "The Book of Machines" narrative, which borrows from Darwin's account of biological evolution to imagine machines following an accelerated evolutionary path of their own, resulting in intelligent machines who are supremely powerful competitors or replacements for, and potential aggressors toward, the human family.

Almost any familiarity with AI or robot representations in popular contemporary science fiction, from *2001*'s HAL9000 to *iRobot*, from Skynet and the *Terminator* to *Westworld*, will spark recognition of this ubiquitous trope. "I, for one, welcome our robot overlords," has been a cultural joke for decades. It is stunning how perfectly Butler anticipates our contemporary fears about AGI in the warnings of the Erewhonian "anti-machinists":

> There is no security . . . against the ultimate development of mechanical consciousness, in the fact of machines possessing little consciousness now . . . Assume for the sake of argument that conscious beings have existed for some twenty million years: see what strides machines have made in the last thousand! . . . No class of beings have in any time past made so rapid a movement forward. Should not that movement be jealously watched, and checked before we find ourselves in a false position and unable to check it?
>
> (190, 194)

Like Musk in 2017, we are warned by Butler's fictional Erewhonian philosopher that unless we act quickly, the machines will soon render us irrelevant and helpless, surpassed and domesticated as we did to the other animals, perhaps lowered to a condition even "worse than monkeys."[3] Even more striking is how the description in "The Book of Machines" mirrors both the capabilities of our current AI tools and their limitations:

> The machine is less versatile; its range of action is narrow; its strength and accuracy in its own sphere are superhuman, but it shows badly in a dilemma; sometimes when its normal action is disturbed, it will lose all head and go from bad to worse like a lunatic in a raging frenzy . . . but here again we are met by the same consideration as before, namely, that the machines are still in their infancy.
>
> (214)

The story of the inevitable replacement of humanity's economic agency will also be familiar:

> ... the machine is brisk and active, when man is weary; it is clear-headed and collected, when the man is stupid and dull; it needs no slumber, when man must sleep or drop; ever at its post, ever ready for work, its alacrity never flags, its patience never gives in; its might is stronger than combined hundreds, and swifter than the flight of birds.
>
> (197)

Here we see a core motif of the AI dream beginning to take shape—an evolutionary process that delivers us from biology, granting immunity to the constraints of organic life and material embodiment. With it comes the other core motif of AI fiction: total supremacy over the living. Butler's fictional scribe admits that the power relation is currently not in favor of the machines: "they owe their very existence and progress to their power of ministering to human wants, and must therefore both now and ever be man's inferiors." But the author warns that even so, "the servant glides by imperceptible approaches into the master."[4] He observes that humans, despite their apparent superiority, are made vulnerable by the dependence on their mechanical servants for basic needs. The obvious solution? Destroy the machines, even at great sacrifice of human comfort, so that man may recover the safety of his dominant station.

Why, and how, could Samuel Butler so perfectly anticipate today's warnings about AGI, almost a century before the first electronic digital computer? Reading his words from our present vantage point makes clear that today's warnings and predictions of "superhuman" machines that turn the tables to enslave us are not actually rooted in the new powers of twentieth- and twenty-first-century computing. With virtually no remainder or substantive novelty, today's dark prognostications reproduce an argument made in 1872 purely on the basis of steam technology. How can we make sense of this? What other shared reality could explain the near-perfect mirroring of today's AGI discourse in "The Book of Machines," if not a common technology?

Ask yourself: What real intelligent beings, not "vapour-engines," have been servants to those at the top of the social hierarchy, forced to perform labor without pay or liberty to pursue their own interests? Which actual beings have threatened that social hierarchy by willfully asserting their own agency and intellect, undermining the rationalization of their servitude and disenfranchisement? Which intelligent beings at the peak of the colonial era in which Butler writes were resisting the yoke, foreshadowing great future struggles for recognition and liberation? Which intelligent beings were, and remain, targets of reactionary oppression, regressive laws, and violence aimed at eliminating the threat and solidifying the existing hierarchy—in Butler's words in his essay "Darwin among the Machines," restoring the "primeval condition" of human dominance rather than letting men sink to the level of an "inferior race"?

The reader may think at this moment that I am comparing steam engines and AI tools with women or with enslaved, marginalized, and colonized peoples. Nothing could be further from the truth. While some have made that comparison as a way to justify academic and policy considerations of the need for "robot rights," a serious comparison of unthinking machines to people is itself a dire moral hazard, as critics Abeba Birhane and Jelle van Dijk have rightly argued.[5] Butler, on the other hand, may not have been immune to that hazard. Perhaps the "machine threat" narrative, now as in Butler's time, is a looking-glass version of a different, very real set of fears that has long enthralled powerful men who see themselves as a natural elite, and as entitled to do whatever it takes to retain and secure that status of unchallenged supremacy.

Think about Alice in her looking-glass dream world, where her disobedient kitten has become the mercurial and powerful Red Queen. What if our own "existential machine threat" narratives are, at least partly, looking-glass projections of their authors' fears of a very different sort of displacement or irrelevance? Now, I think that when Butler wrote *Erewhon,* and "Darwin and the Machines," he really *was* talking about machines, not women or enslaved and colonized peoples. And I believe that when Elon Musk, Stephen Hawking, Sam Altman, and others have talked about AGI, they have been genuinely thinking about AGI, not using code to talk about something else.

But it is difficult to explain the perfect mirroring of the two narratives entirely by means of a common technological cause, given how disparate the basic technological realities they speak to are. And the resonance with colonial narratives of racial supremacy and control is simply too vivid to ignore. It is also why so many readers and scholars were powerfully struck by Ruha Benjamin's account in 2019's *Race After Technology* of how oppressive uses of AI mirror the nineteenth-century Jim Crow Laws that re-entrenched the white oppression of freed Black Americans; a phenomenon she calls the "New Jim Code."

As other researchers have observed, our uses of AI and algorithmic systems, even more so than technologies before them, keep taking on a particular kind of pattern: one that mirrors the colonial-era pattern of resource extraction, power consolidation, hierarchical control, exclusion, and subjugation of marginalized populations.[6] Like the colonial pattern, these uses of AI are rationalized by those who profit from narratives of intellectual supremacy and the inevitable cost of "progress." The commonalities are hard to unsee once they are pointed out. What if seeing them, and acknowledging them, opens a way to liberate ourselves from this pattern?

What if we can, like Alice, wake up from our present looking-glass world in which the most fearful thing the powerful can imagine is their own replacement by intelligent others? What if that would allow us to get a clearer-eyed view and estimation of the real risks and dangers of AI, and how they stack up against the many other, very real existential threats facing humanity today? And what if this awakening would allow our powers of imagination and judgment to craft a better way forward with AI than the path that is currently visible to us?

If the image of AI as an imminent destroyer or supplanter of human dominance and supremacy were only a looking-glass fantasy—a flight of misplaced imagination, serving as a creative template adapted by generations of science fiction writers, then perhaps we could just enjoy it for its imagined possibilities. *iRobot* is a thrill! *The Terminator* is a lot of fun. *The Matrix* is a gas. *Ex Machina* is spellbinding. *Westworld* is mind-blowing (often literally—minds get blown open quite a lot)!

But in fact, the looking-glass illusion these entertaining fictions reflect is being leveraged today by billionaires, lobbyists, and powerful AI companies to directly influence and reshape public policy, scientific research priorities, venture capital investment, and philanthropic giving. It is being placed in service of movements to defer action on real and imminent existential threats—from climate change and ocean acidification to global pandemics and food insecurity—in favor of lavish funding for research on long-term AGI risk, which these powerful interests now claim outstrips all other dangers. The new tech-centered movement of *longtermism,* which grew out of an earlier movement called *effective altruism,* has put forth in many powerful academic, policy, and media circles the idea that AGI presents an incalculably greater threat than even climate change. Many longtermists now argue that the most prudent use of philanthropy and public resources for long-term human benefit is to invest more heavily in AGI risk and AI safety research. Yet it is not the safety of living humans today, or even the next generation, that drives longtermists.

Longtermist arguments often go as follows: because AGI in the far future could theoretically kill or immiserate even more living humans than are alive on the planet now, preventing future AGI from doing this is a more rational use of resources than funding clean energy tech, food security, or public health today. Related claims made by some longtermists include the idea that saving the future requires directing even more of today's resources toward the wealthy in the Global North, rather than sending aid to the most impoverished countries in the Global South, since wealthy people in already disproportionately wealthy regions are better positioned to use their resources to fight these "existential" risks from AGI.

Longtermism is rooted in utilitarian ethics, a brand of moral theory that has long driven effective altruists to the contemporary work of philosopher Peter Singer. Utilitarianism, from its foundation in the nineteenth-century writings of philosophers Jeremy Bentham and John Stuart Mill to the AI-inflected writings of today's longtermists, has always held that our first and highest moral duty is not to root out injustice, or cultivate virtue, or protect human rights and the planet, but rather to maximize the sum total of happiness that will be enjoyed in the future.

Peter Singer is widely admired for his vigorous defense of animal liberation on the grounds of their capacity to suffer, which subtracts from their happiness. He is less well known for his claims (still bitterly recalled by many disability activists) that it is ethical to kill so-called "defective infants" whose potential for future happiness he sees as severely compromised.[7] What longtermists and effective altruists have in common with Singer is the utilitarian view that morality is a happiness optimization problem, a matter of running the numbers. Conveniently for AI enthusiasts, that makes morality just the sort of thing that computers and computer scientists are supposed to be good at!

For a virtue ethicist like me, there are profound dangers and moral errors in any view of morality that separates it from humane feeling, relational bonds, and social context, reducing it to a mere computation of net happiness units. As longtermist Eliezer Yudkowsky readily admits, this kind of fundamentalist utilitarianism licenses one to turn off moral feeling and just "shut up and multiply" happiness sums, even if the action dictated by the calculation feels deeply wrong or shocks the conscience.[8] Some longtermists at Oxford's Future of Humanity Institute (FHI) have used such license to entertain hypotheticals in which the welfare of billions of today's suffering humans is dispassionately sacrificed for the potential to create far larger numbers of happy "digital people" in virtual worlds of the distant future.[9] Others work out longtermist defenses of a related utilitarian view known as the Repugnant Conclusion: the proposal that given the option, it would be better to choose a world packed with people who all suffer so greatly that their lives are "barely worth living" than to choose a far more modest population of ten billion people who can all flourish.[10]

These types of scenarios are often seen as compelling rebuttals to utilitarian logic. When taken instead as plausible moral obligations for humans, such scenarios are, as the philosopher and fierce critic of utilitarianism Bernard Williams famously said, the result of having "one thought too many." Ideas like this did not dominate the early philanthropic thinking of effective altruists. From donating surplus personal wealth to the purchase of mosquito nets and other high-impact aid for the global poor, to investing in public health and climate resilience, much of what effective altruists initially tried to do is aligned with commonsense moral obligations to present and future generations. Even as the effective altruism movement merged with the more speculative AI fantasies of longtermism, many remained committed to fighting the existential threats already here: climate change, biohazards, and nuclear holocaust. As observed by a critic quoted in a 2022 profile on effective altruist and longtermist William MacAskill in *The New Yorker*, "If you read things that [effective altruists] are saying, they sound a lot crazier than what they're actually doing."[11] If the more speculative AGI visions of longtermists were just intellectual side hustles for bored Oxford philosophers and Silicon Valley investors, we might think them harmless.

But today, longtermism is the language of moral thought spoken in a growing number of the wealthiest and most powerful political and industrial circles. Billionaires like Jaan Tallinn and Peter Thiel have been described as avid supporters, while Elon Musk, funder of the Future of Humanity Institute, has cited MacAskill as "a close match for my philosophy."[12] FHI Senior Research Fellow Toby Ord notes in his bio that he has advised "the World Health Organization, the World Bank, the World Economic Forum, the US National Intelligence Council, the UK Prime Minister's Office, Cabinet Office, and Government Office for Science."[13] Ord was quoted to the UN General Assembly in a 2021 speech on climate by former British PM Boris Johnson. Nor do longtermists only seek influence in conservative political circles. FTX crypto founder and longtermist Sam Bankman-Fried was celebrated as a "megadonor" in US Democratic circles before being jailed on fraud charges, while longtermists in the UK formed a group called Labour for the Long Term to influence the Labour Party. Political influence is a core part of longtermist strategy. A 2022 position paper by Oxford's Global Priorities Institute outlined their political case for *institutional longtermism*, the view that far-future considerations should dominate public policy and even our choice of political institutions, in contrast with longtermism as a personal philosophy for setting one's own philanthropic priorities.[14]

The AGI looking-glass illusion plays a key role in driving this philosophy in a dangerous new direction known as "strong longtermism." On this view, far future threats, even highly speculative ones like AGI, simply dwarf even the most urgent moral claims of presently living or soon-to-be-born humans. As MacAskill and his coauthor Hilary Greaves argued in 2019, "For the purposes of evaluating actions, we can in the first instance often simply ignore all the effects contained in the first 100 (or even 1000) years, focusing primarily on the further-future effects. Short-run effects act as little more than tie-breakers."[15] In his 2020 book *The Precipice*, philosopher Toby Ord argued that the odds of human extinction are magnified far more by the risk of a malevolent or misguided artificial superintelligence than by all other existential risks combined, including nuclear holocaust, bioterrorism, and climate-driven biodiversity collapse.[16]

After all, environmental calamities could rapidly make the planet unlivable for the vast majority of humans but would probably enable pockets of survivors to endure. AGI, Ord claims, could easily kill us all, perhaps by blackmailing or manipulating world leaders to initiate a nuclear holocaust. Ord does not claim to know that AGI *will* emerge, or that if it does, that AGI will take such actions. He only says that it is theoretically possible. Yet using the kind of tortured math that longtermists love, he claims that even a near-zero possibility of total catastrophe in five hundred years is more urgent than a strong likelihood of our near-total destruction in fifty.

Let's stop and take a breath. First, there is an aspect of this AGI scenario that might provoke some readers to laughter, so let's address that. Our best AI systems today can solve some pretty complicated optimization and prediction problems, but they can't think their way out of a paper bag. Today's AI systems still fail, spectacularly and often. They are hard to build and hard to keep working well. They are subject to model drift, the rapid performance decay that happens when a model is exposed to real-world data that diverges from its training data. We've seen that even generative AI tools, the latest to be held up as the key to AGI, routinely generate falsehoods mixed imperceptibly with facts—in fact, the more powerful these models are, the more they seem to fabricate.

Moreover, AI tools only work at all when we labor tirelessly to feed them the carefully cleaned and curated data they need, tune their delicate hyperparameters, and keep them operating in familiar, well-constrained environments. Not to mention the immense quantities of water, energy, and mined rare earth minerals needed for the graphics cards and server farms that power AI, which humans must provide. For many AI experts who live with these struggles, imagining a future version of today's AI taking over the world from humans can remind one of the dinner scene in the 2008 film *Forgetting Sarah Marshall,* where the characters crack up laughing over the premise of a cheap horror film centered on a murderous mobile phone. As one character jokes, "Why couldn't you just take the battery out of the phone? . . . Yeah, we've won."

But of course, those who worry about AGI existential risk imagine scenarios where humans have already given a malevolent AI system control of our power grid, and our water supply, and the mineral mines, and the chip factories, and everything else that AI needs to exist. Of course, we could simply *not do that.* This doesn't seem hard. But that does not mean that AI poses no grave safety risks. In truth, we actually have more to fear from an AI system that is *not* truly intelligent, that doesn't have a clue what it is doing or why, but has nonetheless been handed control of the critical systems on which we depend, in the name of "efficiency" and "innovation." This would be immensely foolish, as many AI tools are prone to dangerous acts of "reward hacking." That is, while it can't alter its "reward function" or programmed objective, a machine learning system can often calculate a more efficient, surprising, and highly undesirable path to obtaining that goal. This is an old trope of science fiction, where AI tools that we program to protect human life at all costs might just imprison us to guarantee our safety. Real-life examples of AI "reward hacking," while far less dramatic, are legion. They highlight that predicting how mindless AI tools might go about achieving the goals we set them is a serious problem. In the words of cybernetics pioneer Norbert Wiener, "To turn a machine off effectively, we must be in possession of information as to whether the danger point has come. The mere fact that we have made the machine does not guarantee that we shall have the proper information to do this."[17]

There is, in fact, good reason to dedicate part of AI research today to AI safety in order to ensure that these kinds of human design choices are not made. Let's not give AI tools any opportunity to reward-hack the power grid, or our water systems, or the nuclear launch codes. These are commonsense protocols that should be at the center of national security policy, international regulation, and AI development standards. But we should ask why contemporary AI narratives so often fixate on the prospect of AI *wanting* to destroy us, of seeking to enslave or exterminate humanity rather than harming us as mindlessly and accidentally as a parked vehicle left in gear. What is behind this expectation of genocidal intent? An obvious answer is that we have seen such intentions in humans, who are intelligent. But this is a poor argument. Why should we expect genocidal intentions to be causally linked to intelligence? That's a pretty curious and disturbing assumption, one with no obvious grounding in the science of intelligent life. It's an assumption that tells us more about ourselves than it does about technology, or about intelligence.

In fact, if AGI did one day emerge, it would likely be a very different kind of intelligence from ours, not doomed to replicate our mistakes or mirror our weaknesses. As philosopher Peter Godfrey-Smith discusses in his book *Other Minds,* as did artist James Bridle in *Ways of Being,* we already share this planet with multiple other intelligences, from octopuses, to elephants, to dolphins, to crows—and possibly even to underground networks of plants, if we are willing to expand our anthropocentric standards far enough. Each of these other intelligences reveals itself in radically different ways. As Bridle points out, one reason that it took scientists so long to discover the intelligence of non-human animals is that for most of the last century, we measured their intellect with tests designed only for humanly embodied intelligence. These were tests that could be mastered by animals with two legs, thumbs, and eyes that looked straight ahead. When we started investigating what novel problems animals could solve with eyes that looked up instead of forward, or using eight legs instead of two, or with beaks and tails instead of thumbs, guess what happened? We suddenly started seeing a lot more intelligence!

Let's imagine the different paths by which AGI might be pursued. One involves experimenting with many ways for AI systems to learn about their environment, including biologically inspired system designs that borrow from the rich diversity of living intelligences. Research of this kind is already carried out under various labels, from bio-inspired computing to artificial life and biomimicry. While this kind of AI research seeks to mirror certain aspects of living systems, it has the potential to enable something far more ambitious than the parroting of observed human language or behavioral patterns seen in most commercial AI systems today.

In his 2022 position paper "A Path Toward Autonomous Machine Intelligence," Yann LeCun challenged dominant machine learning trends, arguing that neither large language models like GPT nor reinforcement learning models such as AlphaZero will lead to AGI. Instead, he proposed an approach based on unsupervised learning of a commonsense model of the world through a hierarchy of world representations, arranged in a cognitive architecture that enables the agent to think using a single "configurable world model engine." He compares this hypothetical "engine" to the capabilities of the animal prefrontal cortex. To LeCun, our excitement about AI tools that mirror human language skills obscures the fact that intelligence evolved in many animals without, or prior to, complex language systems. To put the point simply: if our nearest primate ancestors were already intelligent, and their intelligence laid the foundations for ours, then starting with language seems like trying to frost the cake before you've baked it.

But that is exactly the approach taken by today's most heavily funded AI research programs and commercial AI ventures: building larger and larger AI mirrors focused narrowly on human performance, especially linguistic performances, which just so happen to make up the lion's share of cheap data scrapable from the Internet. On this path, we feed AI systems as much human-generated data as we can, training them to learn and mimic all the ways that we behave and speak, to match all the patterns that link our perceptions to our goals and desires, and to replicate all the predictable impulses and preferences that we ourselves demonstrate. Stanford economist Erik Brynjolfsson has called the aim of this type of AI *human-like artificial intelligence* (HLAI) to distinguish it from the broader frame of AGI, which could include machine intelligences that are not like us at all.

Why are we trying to build AI to mirror ourselves? In part, because we can. That is, we already have what we need to make it work. Data and processing power are the two computing resources upon which today's approach depends. Large, well-funded AI companies have been able to get their hands on a lot more of both in this century, thanks to the exponential growth in computing power described by Moore's Law and the equally rapid growth in human-generated data—much of it delivered by social media and Internet platforms. Recall what I said in Chapter 2, that data are the incident "light" of these new AI mirrors. The answer to why so many AI researchers are now committed to this approach is a version of the punchline to the old joke: Why is the drunk man searching for his lost car keys under the lamppost? Because that's where the *light* is.

Still, a lucky drunk might find his keys there! But even if our present path did somehow lead us to "human-like" AGI (setting aside that our AI-training data miserably fail to represent all humans), consider this plainly obvious fact: we already have people to be people. Some countries do face temporary labor deficits due to aging populations, particularly when paired with rigid and punitive immigration policies, but globally, people are not in short supply. Even if it were feasible, creating digital versions of humans seems, well, redundant. The obvious exceptions involve tasks that very few people are capable of performing, or those tasks that most people would be happier with machines doing—what AI researchers call the 3Ds: dull, dirty, and dangerous labor. But even *if* many such tasks should be fully automated (a question which invites complex moral, political, and economic deliberation), narrow AI should do just fine for these applications. It doesn't explain why we should build AI mirrors to be our doppelgangers.

Of course, there are obvious commercial advantages to creating artificial people because, as Samuel Butler pointed out, they don't need sleep or holidays. They don't ask for promotions, and they don't take family leave. But what point, then, is there in trying to make these mechanical mirrors into people at all? What today's AI research trajectory seems to be aiming at is a dream that should deeply disturb us: silicon persons who perfectly mimic or exceed humans' flexible, adaptable capability to perform economically productive tasks—writing movie scripts, reporting news, predicting stock prices, serving coffee, designing posters, driving trucks—but with every other capability of a human person left behind, unreflected and unmade.

This is exactly what's so dangerous about the rhetoric of "superhuman AI" commonly invoked by today's AI leaders, from Yoshua Bengio, to Yann LeCun, to Geoffrey Hinton and Sam Altman. They describe as "superhuman" those AI systems that will be better than we are at calculation, prediction, modeling, production, and problem-solving. Are these the capabilities that make us human? What metric for the value of a human life do we endorse when we claim that a machine that does nothing else but compute and produce has earned the title of *super*human? Are you solving a problem when you spontaneously pat your child's head in affection? Are you optimizing your production yield when you do a cartwheel in the grass? Are you calculating your most efficient path when you stop in the street to check on a stranger in distress?

The fearsome qualities of superhuman AI described by many of today's AI leaders more closely resemble those of a *virus* than any human being. Like a virus, they envision superhuman AI ruthlessly optimizing for its own survival and replication, without consideration of any interests beyond its own. 'Superhuman' AI will mindlessly exploit our bodies and manipulate our behaviors, hacking our defenses to maximize its productive output—all the while, utterly indifferent to the meaning of killing us. If this is what you think it is to be superhuman—that is, just like a human, only more so—I think you have completely forgotten yourself.

Brynjolfsson notes that this "Turing Trap"—building AI specifically to mirror economically productive human performances—carries grave risks. The first is the disempowering of most humans on a planet where power is already unjustly concentrated in the hands of a few. He observes that when a human has personal possession of the knowledge needed to perform a task, "that person can command not only higher income but also a voice in decision-making."[18] But when that personal knowledge has been copied onto a digital artifact that anyone can replicate, the economic and political influence of the person is greatly diminished. As Brynjolfsson says, "When useful knowledge is inalienably locked in human brains, so too is the power it confers. But when it is made alienable, it enables greater concentration of decision-making and power."[19]

We also stand to gain more, even in purely economic terms, from building AI systems that *augment* human capacities rather than mirror them. Brynjolfsson points out that AI tools often yield the highest performance when they are paired with a human agent in a complementary way: when we build them to amplify or strengthen one component of a human performance rather than replace it altogether. For example, an artificial agent built to function as a virtual therapist chatbot yields mediocre and inconsistent results. As we saw earlier with the chatbot that endorsed suicidal thoughts, the worst-case scenario can be catastrophic.

But one company, Crisis Text Line, has used AI to rapidly match human volunteer therapists with people calling an emergency crisis hotline. This is a task that a very narrow AI tool can perform effectively and nearly instantaneously, all without a whit of "human-like" intelligence. It's also a task that doesn't replace a human ability—no human could do this kind of matching in real time without delaying lifesaving help. But with an AI tool in their task loop, matching their skills to others' needs, the human therapists are augmented; they can be more effective than they were on their own.

Using AI in this way doesn't only make us better at what we already do. It can also help us do new things that were formerly beyond our capacity. And this is going to be of vital importance in this century and the next, as climate change, food insecurity, and global pandemics will require us to do many new things to survive and protect life on the planet. As Brynjolfsson notes, "inventing tools that augment the process of invention itself promises to expand not only our collective abilities, but to accelerate the rate of expansion of those abilities."[20] He observes that a turn from automation to augmentation can also counter the growing "tech backlash," largely driven by justifiable public reactions to tech that increasingly exploits or replaces rather than empowers people. This backlash, however warranted, threatens public confidence in innovation, and undermines our collective will to develop and embrace the new science and technology that can boost our capacity to meet the existential challenges we face.

There is another risk of the Turing Trap that Brynjolfsson does not mention, however. It's not a speculative future harm, but one that's already happening on a global scale. This is the use of human-like AI mirrors to enforce what philosopher of technology Langdon Winner in 1977 called "reverse adaptation."[21] Here, instead of machines being designed to support human ends, humans are enlisted to do whatever is necessary to augment and adapt to the machines' abilities. In this way humans are rewarded for behaving more and more like robots. It is what happens whenever AI tools instruct Amazon warehouse workers to bend their bodies at the "optimal" angle for box-picking speed, ignoring the human worker's unique embodied situation (the sore right shoulder or the tricky knee) and demanding that they move only as the "optimal" box-picker encoded in the machine mirror would.

It's what happens when AI monitoring tools used in schools in China and elsewhere track students' eye and facial movements, ranking children on how well they match the predicted patterns of "optimal" classroom emotion and attention to their lessons. It's what happens when an AI "ride-along" monitor installed in the cabs of long-haul truckers locks out the driver's skillful judgment and instead instructs them on the timing and location of every turn, rest stop and meal break, turning a skilled, experienced human into a meatbot behind the wheel, passively mirroring the machine's "optimal" choices.[22] It's what happens when AI cosmetic filters lead young people to surgically alter their faces to match the distortions they see in the AI mirror.

In a profoundly ironic turn, this was also the plan for the data collected by nonprofit Crisis Text Line, when it packaged anonymized data from the sensitive conversations between their human volunteers and callers for sale to its for-profit spinoff Loris.ai.[23] I say, "was the plan" because a loud public outcry from the volunteers and a global community of data and AI ethicists stopped it. The troves of data that Crisis Text Line had amassed—extracted from deeply personal and charitable acts of humans helping other humans in crisis—was going to be used by Loris.ai to power its customer service optimization AI tool, which monitors, instructs and scores human call center workers. The Loris.ai website says its AI improves the efficiency and 'empathy' of individual agents by steering their responses into "more predictable" patterns that "make every agent, your best agent." Is this how humans acquire empathy—by parroting a machine-supplied pattern?

We can build AI to mirror and systematically replace human intelligence—and empathy—with a machine-molded facsimile.

Or we can build AI to genuinely assist and liberate our humane potential. You might ask, "why not both?" As Brynjolfsson points out, our short-sighted economic and political incentives have already heavily stacked the deck toward the first strategy. Unlike the ruthless optimization promised by the path of mechanized agency, the second option allows AI to increase the power and agency of beings like us. Beings who can say no, ask difficult questions, or point out inconvenient truths. Beings who can value things like justice and liberty more than efficiency or profit. Beings for whom intelligence is physically inseparable from the need to sleep, joke, mess around, doodle, daydream, and blow off steam. If we stay on the path with AI we are traveling now, that will not be the kind of intelligence that guides our futures.

More likely we will get stuck in a world where our complex human agency, spontaneity and intellect are increasingly cramped and restrained by unthinking AI mirrors, wielded by powerful corporations and government powers who demand that our actions reflect back to these dead mirrors a denatured, reduced, but "optimized" version of what they took from us. This will also be what drives AGI development. Even if AGI cannot arise by this path, this model of "superhuman" intelligence will nevertheless define the kind of technology that we and our children will live with, and it will become the main measure of their humanity. I hope the examples above give you ample reason to question the desirability, or even the moral acceptability, of that future.

The growing trend of reverse adaptation to AI explains why our lives will be worse if we don't alter the current trajectory of AI development. But the examples above don't capture the full sadness of such a failure of humanity's most meaningful existential task: what Ortega called our autofabrication. Today's AI mirrors cannot guide our future without us surrendering every hope of making ourselves more than what we have already been. Recall the earlier chapters' description of today's AI tools as mirrors pointing backwards, narrowly reflecting only the dominant statistical patterns in our data history. Such mirrors, when used not as reflections of the past but as windows into our future, serve as straitjackets on our moral, intellectual, cultural, and political imagination. They project the statistical curve of history into the still-open future, and by rebranding that reflection as a *prediction,* they profoundly restrict our sense of the future's possibilities.

Popular narratives of genocidal machine intelligence narrow that view further. Today's most widely consumed visions of AGI or "true" artificial intelligence strike me as profoundly stale and uninspired. The nineteenth-century colonial and industrial anxieties of machine supremacy that drove Butler's *Erewhon,* reconstituted in the dystopias of *The Terminator, The Matrix,* and *Westworld,* now seep from pop culture into policy and politics via the dire warnings of today's longtermists. Yet even as cultural mirrors, the reflections these give are shockingly poor. Instead of the virtues of wisdom, playful curiosity, and imagination that animate the history of human intellect, today's dominant image of AI only mirrors the aspects of that legacy that we most desperately need to surpass and transcend.

Today's AGI visions of callous optimizing machines, using humans as mere fodder, mirror the nineteenth-century factory owner's shortsighted impulse to use up every material resource, and every ounce of his workers' own vitality and health, without remainder. Visions of AGI overlords that cruelly turn the tables on their former masters mirror the same zero-sum games of petty dominance and retribution that today drain our own lives of joy. These visions of the "superhuman" reflect an utterly barren ideal of human life, one constantly optimized and optimizing for a singular, predefined goal, but never feeling, playing, or aimlessly exploring. It reflects our vain, futile desires for control and supremacy, but leaves beyond the penumbra our compassion, our curiosity, our impulse to care for one another, and our capacity for solidarity and love. It reveals a vision of calculating intelligence that is unimaginably cold, flat, and above all, dull.

Consider the American TV series *Westworld*. The story arc of its first few seasons told an emotionally compelling story of how sentience and free will might arise in a mechanical being through memories of loss. But in season four (spoiler alert!), we finally got to see what kind of world the now-conscious artificial "hosts" built after freeing themselves from the violent yoke of human oppression. Since the first season, viewers had awaited the fulfillment of host Maeve's equally grand and profane promise of the gift of autofabrication: "In the new world, you can be whoever the fuck you want." So, after all the possibilities we imagined the liberated AI hosts could explore, what new world did they build for themselves?

They built the most boring, uninspired, and uninspiring world imaginable. The final season shows us dull, gray replicas of Earth cities where humans are forced to run on fixed software loops, pointlessly tortured like flies by the freely roaming hosts. The hosts are now just mirror images of their creators' cruelty; as one notes: "we are reflections of the people who made us." The captives have become the captors. Yawn. Just like AI mirrors, the hosts' reflections of our humanity aren't even as rich as the minds of their inventors. We don't see the hosts doing anything more interesting than what the most tiresome human alive today would do.

In one notable scene, a host has a human couple trapped at a dinner table in a fine dining restaurant. Even though the tension in their bodies shows a clear desire to flee, the couple are forced by the host's AI technology to remain seated and engage him with verbal pleasantries as he narrates his power over them, smirking over his fine single malt. It's a perfect parody. We can imagine any billionaire today doing the same with his trapped underlings, recounting the fine details of his latest yacht purchase, boardroom triumph, or pointless trophy—some endangered animal he's recently eaten, shot, or made into a helpless pet—while the people who depend on him for a living or their status just sit and grin with forced rictus smiles. The only difference in the *Westworld* scene is that the people at the table are themselves the pointless trophies.

This "new world" is so boring that the hosts, given the chance to live forever, or even to upload into new mechanical bodies less confined by human physiology, instead start killing themselves from ennui. They'd rather die than remain in the world created by their supposedly uber-intelligent leader. It's the world they have already lived in from the other side, and it is somehow even less interesting now that they hold its reins. Repeatedly we are told that the hosts' human blueprint doomed them to this fate. They are reflections of us, so this is all they can be. "Violent delights have violent ends," and all that. There's a faint suggestion that the hosts who escaped into the virtual world of the Sublime in a previous season might be doing something more interesting with their lives, but we don't get any sense of it. Why not? What does it say that when we imagine beings built in our image, the only purpose we can think of for their intelligence, the only task in which we imagine them seeking pleasure, is violent domination and control? Is that really what intelligence is?

Why do so few depictions of AGI show us a superhuman intelligence that laughs more than we do? Where are the intelligent machines not on a mission, but mastering being silly, goofing off, exploring, playing? Most highly intelligent creatures do a lot of this! What do our imagined superintelligences ever read, write, sketch, or sing? What wild new music do they make? What living spaces do they build, other than white plastic hyper-modernist exaggerations of our own blandest aesthetic? What games do they invent that are not just our worst ideas turned up to 11?

Every existing intelligent being that we actually know of shows signs of some or all of these broader, richer capacities. Some of this planet's intelligent creatures assuredly have capacities that we cannot even describe. But AGI? All the most powerful, influential, popular voices in the AGI conversation can imagine is a ruthless killing or producing machine: one whose highest impulse, in the oft-cited example from longtermist Nick Bostrom's 2014 book *Superintelligence,* might well be to convert all the matter in the universe into a giant paperclip factory.

Some less widely known fictional visions of AGI do far better. But those aren't the visions driving policy discussions of AGI futures today or shaping mainstream AGI research. They certainly don't guide the alarmist predictions of the longtermists. In a 2020 podcast on existential risk, *Precipice* author Toby Ord insisted that AGI will "dethrone" humans, knocking us out of our "privileged position" of control over all earthly life.[24] AGI will demand that it "call the shots." It will regard other intelligent beings as no more than resources, "just there as something for them." In this looking-glass vision, AGI "spreads out over the whole world so it can affect everyone." Are these visions of AGI a prediction? Or are they a *confession*?

Like Samuel Butler and the writers of *Westworld,* longtermists routinely depict an artificially intelligent being as a mind that could only see it as rational to replace, enslave, or murder humans, and to use up all our resources. Why? Because that's what the strongest, smartest, or most powerful of us inevitably do, and have always done? If this is what we think, we are telling on ourselves. When we are absorbed in our most popular AGI fantasies, we are, like Alice, lost in the looking-glass world. We imagine that we are looking through a window into a radically new and dystopian future with

AI, when we are really stuck in the past, dully gazing backward. And not even our entire past, and not everyone's past—instead our AGI fantasies reveal only the narrow slice of human experience told in our stories of conquest, domination, and empire.

To conflate that story with the story of human intelligence is either horrifyingly deluded or willfully perverse. Even if we wrongly equate intelligence with calculative self-interest, no exemplar of human intelligence I can think of has wisely chosen a life of humorless, friendless misery, unchecked rapaciousness, unquenchable ego, and a relentless desire to punish. In the interest of not inviting lawyers' fees, I'll just ask the reader to think of the humans best fitting the latter description. Do they look like they've got it all figured out? Are they happy? Making the smartest choices? Our AGI visions are conflating intelligence with the abuse of unearned power, which is a pretty revealing mistake, and one we make all too often. For it is manifestly not the case that the most intelligent creatures—even among the human population—tend to choose the abuse of power as their objective function. Those who do choose it, and keep choosing it, tend to live lives of anger, alienation, and loneliness that would horrify even the most strident utilitarian.

Of course, it isn't impossible that our efforts to create a truly artificially intelligent being would instead accidentally produce a malevolent, genocidal exterminator. Longtermists have argued that this possibility, however unlikely, warrants the undertaking of massive preventive efforts today. Certainly, work in AI safety research should continue. The question is whether it warrants draining vital government funding and philanthropic resources from our collective response to the already real existential catastrophes currently knocking at the door.

If we do not make an all-hands-on-deck effort to salvage human civilization from runaway climate change, ocean acidification, biodiversity loss, and resulting global ecological collapse of the food chain, there will be no university labs or well-funded startups left to build Skynet—or anything else. Steering limited public and philanthropic attention and resources away from these crises in order to fight a malevolent AGI in our looking-glass future is as rational as sending away the fire crew who are rushing to extinguish your home's burning roof, simply because you want to install a burglar alarm.

Longtermist narratives that present this looking-glass AGI as the most urgent threat to our humanity suffer from a dual failure of moral imagination. First, they fail in not being able to imagine enough. They have a curious tunnel vision of what AGI would be like. As we've seen, mainstream AGI fantasies manage to exclude almost everything we know about the behavior of intelligent creatures. What these fantasies reflect is not a plausible vision of intelligence at all, but a digital bogeyman—a silicon golem of our own worst failings. If we think that's where our current AI efforts are now heading, then we are already doing it wrong. As critics observed of the 2023 flurry of open letters on existential risk signed by many powerful AI industry leaders: If you really think you might be building a superpowerful human extinction machine, then rather than ask governments and philanthropists for piles of money to build stronger handcuffs for your giant human extinction machine, why don't you stop and go build something better?

The second failure is the inability to properly envision the existential moral stakes of the immediate perils that are now coiling around us. Longtermists who quake at the thought of robot overlords will speak of catastrophic climate change with the detachment of an actuary. In the 2020 podcast, Toby Ord and the host considered the extreme prospect of global warming by 10–20 degrees, in which the habitable regions of Earth shrink by two-thirds, or more. Ord opined on the compatibility of this scenario with continued human flourishing: "It'd still be much more habitable, than, say, Mars . . . there would still be many places one could be."

Who is this "one," or ones, in whose circumstances Ord asks us to find comfort? And where are the rest? Eight billion people currently struggle for survival on the habitable land and with the potable water we have. By 2050, there will be ten billion, and far less of both. In a future with only a third of today's habitable land (but still places that "one could be"), what has happened to the *others*? Ord, seemingly alert to the chance that he might be seen as callous, noted, "It'd be very bad, just to be clear to the audience." But he never actually spoke of the missing ones, or their demise. He did not describe what it would be like for the ones who endure to see billions vanish. There is no mention of the waves of climate refugees wasting away in armed camps, trading their bodies for food, drowning in floods on parched earth, their children buried in deserts that were once gardens. The podcast was not about them. It was about running the numbers. The others just disappear from the story. Or rather, we must forget that they were ever in it.

This silence, this vision averted from what a failure to prevent climate catastrophe will actually mean, embodies the second failure of moral imagination—a virtue that differs profoundly from the imagination of the child. The imagination that transports Alice into the looking-glass world is her untutored freedom to bring anything to her mind that she would like to see, and to transform anything she does not want to see into something else. Moral imagination, on the other hand, is the habit of bringing to mind those worldly possibilities that we can and must make *real*. Equally, it is a habitual refusal to avert the mind's eye from the harder potentials to contemplate, the ones we would rather not see, the ones that make us squirm, cry, or ache. Moral imagination marries our creative vision to the virtues of responsibility, wisdom, compassion, and courage.

No starker contrast could be drawn between the narrow abstractions of Bayesian probability calculus that dominate longtermist narratives and the moral imagination that animates the opening pages of Kim Stanley Robinson's 2020 novel *The Ministry for the Future*. In a sense, the novel aims at the very same end as the longtermists—illuminating our place on the precipice of existential risk and urging us to collective action to avert it. Though nominally fictional, *The Ministry for the Future* is laden with many sober ruminations on real-world policy, economics, and science of climate risk mitigation. But it does not begin in the places where "one could still be." Instead, it begins with the *others*. It begins with a vision of local, embodied suffering in a boiling world that is so vivid, so wrenching, so unprecedented and yet so imminent for us that many a reader must close the pages and remember to breathe. In the novel, it is not the safety of the last, but the loss of the first, and the suffering of one who is unlucky enough to be left as their witness, that drives everything which follows. For those whom unsurvivable heat and flame have already forced into the sea of Rhodes, Sicily, or Maui, the book's opening scene will no longer be fiction.

We are in the midst of an accelerating sixth mass extinction, on a planet that we are making inhospitable to humans faster than even the most worried scientists had anticipated, with less than a decade left to prevent irretrievable loss. We cannot afford to get lost in looking-glass futures where threatening mechanical apparitions of our own making await. But we also cannot do without the virtue of imagination matured by practical wisdom. In fact, we need the power of imagination more than ever. And AI mirrors may be of help in cultivating that virtue; their narrow beams of vision might be used to augment and widen ours.

For example, can large pretrained AI models function as a new kind of digital muse? Many artists have found great joy in playing with tools like DALL-E and Midjourney to generate new visual ideas and spark their own creative potential. And art has always been one of the best ways to enlarge our moral, scientific, and political imagination. Yet others see the latest AI tools as aesthetic parasites, feeding on massive data sets of artworks created by humans who receive no credit or compensation for AI-generated outputs. Fears that human artistic potential will be replaced by cheaper, shallower reflections from AI mirrors were first awakened when Jason M. Allen, the winner of a 2022 fine art contest in the United States, revealed that he had won his category with a digital painting generated by feeding a text prompt to Midjourney. These fears deepened in the wake of the 2023 Writers' Guild of America strike, which prompted some studios to consider the use of generative AI as a new form of "scab" labor.[25]

Changing the current economic and legal incentives could enable AI mirrors to augment rather than extinguish, cheapen, or replace human imagination. But for that to happen, our raw power of imagination must be steered by mature moral ambition and practical wisdom: the virtue of phrónēsis. A curious fact about this virtue relates to our driving metaphor. After the shift from Greek to Roman culture, the virtue of phrónēsis reappears as *prudentia,* one of the four cardinal virtues of Roman thought, along with justice, courage, and temperance. And when the virtue is personified in medieval, Renaissance, and Victorian art, Prudentia almost always holds a mirror. Why?

We translate *prudentia* in English as "prudence." In this language, "prudence" is often associated with a Victorian attitude of caution or modesty that is absent from the older meanings of phrónēsis or prudentia. For this reason, most philosophers tend to prefer the term "practical wisdom" when speaking of this virtue. But still—why does the personification of practical wisdom as Prudence hold a mirror? She also usually holds in her other hand a snake, or sea creature. While there are no definitive historical explanations, the snake may represent our worldly troubles—the thorny, tricky challenges that practical wisdom helps us to confront.

Prudence's mirror is likewise open to multiple interpretations. One of them is that practical wisdom requires a clear sight of things, the ability to see them as they are. If mirrors are, like Plath's "four-cornered little god," able to show us reality "unmisted" by our own preconceptions and expectations, then this would be an aid to practical wisdom. Mirrors are also tools of self-reflection, which always accompanies wisdom. Yet a third interpretation, which explains the curious way Prudence's mirror is often aimed over her shoulder, is that a mirror allows us to see *behind* us. An essential part of wisdom is to know our past and how it bears upon the present, to understand how history's patterns shape our present dilemmas and future possibilities. Prudence's mirror represents that knowledge.

Supporting this interpretation is another common motif in depictions of Prudence. In works such as that of the Renaissance painter Titian, Prudence has a second face—on the other side of her head, facing backward. The most plausible interpretation is that, as Aristotle noted, practical wisdom is born from—and continues to learn from—past experience. But learning from experience is not the same as repeating it. Experience teaches us how our responses to our past worldly troubles panned out. When we reflect on experience, we can double down on successful strategies and imaginatively adapt them to new circumstances, but we can also abandon strategies that have repeatedly failed us. The mirror's reflection is not wisdom; it is a tool wisdom uses.

As we have seen, if AI's backward-facing mirrors are built with the right kind of data, we can use them to extract subtle patterns from our past. They can enable us to learn more from our history, and help us avoid repeating our worst mistakes. But there is much more in our past than what our digital data sets capture. And when we do catch sight of our past in the AI mirror, it is essential that we do not mistake those patterns for destiny or allow them to become self-fulfilling prophecy. After all, Prudence uses her mirror to walk more wisely *forward*. If we mindlessly replicate what we see in the mirror of history, we will never build upon that knowledge, never be free to try new and better approaches. Today, we stand at a critical moment, facing a choice: Shall we make real once again the ghostly images that appear in our digital mirrors of empire, extraction, and domination? Or shall we imagine something else to make from ourselves, something new?

We must not take the mirror of human history as a doomed blueprint for AGI, or for ourselves. But to do otherwise, we first need to challenge the way we use today's AI tools. Increasingly they function in our hands not as mirrors of prudence, but as engines of automated decision-making that engrave ever deeper into the world the patterns of our past failures. We need the courage to imagine more than the zero-sum games of dominance and exploitation. These were never the hallmarks of intelligent life. These roads have ended for us; they cannot be walked further, and what lies beyond them is desolation. We must seek other roads into the future, charted with the aid of our best technologies—but most of all, with the virtues of moral imagination, courage, and shared wisdom. This is our new world; not a looking-glass projection of the old.

Chapter 4

The Thoughts the Civilized Keep

> The painter who draws merely by practice and by eye,
> without any reason, is like a mirror
> which copies every thing placed in front of it
> without being conscious of their existence.
>
> —Leonardo da Vinci, *Libro di Pittura*
> "Book on Painting" (1513–1516 AD)

In 1911, the philosopher Alfred North Whitehead wrote the following in his book *An Introduction to Mathematics*:

> It is a profoundly erroneous truism . . . that we should cultivate the habit of thinking of what we are doing. The precise opposite is the case. Civilization advances by extending the number of important operations which we can perform without thinking about them.
>
> (46)

Of course, Whitehead wasn't talking about AI systems, or their power to extend the number of important operations we can perform without thinking. He was talking about something very closely related to AI, however: the use of mathematical techniques of formalization and symbol manipulation to automate certain kinds of problem-solving. Whitehead uses the example of algebraic formulas that represent basic laws of commutation and association using terms such as *x*, *y*, and *z*, and symbols such as +, *()*, and =. He points out that these formulas allow certain kinds of reasoning to be performed by us "almost mechanically," without having to call into play "the higher faculties of the brain."[1] He's talking about the power of algorithms: rules for manipulating symbols to solve problems.

Whitehead did not want algorithms to entirely liberate humanity from the burdens of thinking. His point was that thinking is, for humans, a costly business: "Operations of thought are like cavalry charges in a battle—they are strictly limited in number, they require fresh horses, and must only be made in decisive moments."[2] The more of our mental labor we do mechanically through rote computation, the more cognitive energy we keep in reserve for when genuine thought is needed.

But Whitehead did not anticipate a world in which machine learning tools might promise to do even the work of the cavalry, and for a far lower cost. That is, he did not anticipate a world in which those "higher faculties" of the human brain would have to compete with computational systems that produce endless mirror images of human thought faster and more cheaply than we can. He did not envision the design of mechanical decision-makers that promise to automate scientific hypothesis generation, artistic and literary creation, architectural design, military strategy, national policy, life-and-death moral decisions, and even acts of religious benediction.[3] Today, AI mirrors are designed to automate the most consequential and complex forms of thinking. What does this mean for our shared humanity? A century after Whitehead celebrated algorithmic automation's vital role in driving the progress of human civilization, we must pose the question that he did not see the need to ask. What thoughts do the civilized keep?

Much of today's media obsession with AI and the "power of algorithms" obscures the fact that algorithms, as finite series of steps for generating solutions to a given cognitive problem, are nothing new. Both the term (which first appears in late-medieval Latin, named for the ninth-century mathematician al-Khwārizmī) and its referents (which include the earliest mathematical procedures for counting, addition, and division) long predate modern computing. Yet modern computing has invested today's digital algorithms, especially those embedded in AI-driven automated decision systems, with vastly expanded social power.

Today, increasingly sophisticated algorithms determine and shape what we read and write, what we watch and hear online, who we are invited to meet or date, what medical treatments we are advised to undergo, who will hire us, how the justice system will treat us, and where we are allowed to live. Further social constraints and influences from AI and decision-system algorithms are projected in almost every sphere of culture, government, and commerce. Since new mechanisms of social power are always philosophically significant, we should not be surprised to find that these developments raise new and urgent political, epistemological, and moral questions. Among them are questions about the opacity of the social mechanisms now dependent on these new computing techniques.

For it has become increasingly challenging to understand exactly when, how, or by whose authority these algorithms produce their profound influences on our lives. This *algorithmic opacity* or lack of transparency in a "Black-Box Society," to borrow the title of Frank Pasquale's excellent 2015 book on this subject, raises profound ethical questions about justice, power, inequality, bias, freedom, and democratic values in an AI-driven world. The problem of algorithmic opacity is especially complex given its multiple and overlapping causes: proprietary technology, poorly labeled and curated data sets, the growing gap between the speed of machine and human cognition, and the inherently uninterpretable and unpredictable behavior of many machine learning processes. The latter can prevent even AI programmers from fully understanding the internal operations of the AI system they designed, learning the cause of any given output, or assessing its reliability in complex interactions with other social and computational systems.

What makes this opacity so hard to remove? We have seen how today's AI systems mirror and magnify the statistical patterns they extract from the vast data pools used to train each machine learning model. The most powerful AI mirrors today can extract patterns that humans looking at the data could never find. This is partly because of the greater computational speed of their processing, but also because of the sheer size and complexity of models of this type, which belong to a class of machine learning called "deep" learning. Deep-learning models are built with a highly complex network structure composed of multiple mathematical layers. The structures of these layers and the connections between them are defined by variables called parameters or weights. An early example of a large language model, Google's Pathways Language Model (PaLM), had 540 billion of these variables. Others are now trained with more than a *trillion*.

By feeding computations across its many internal layers forwards and backwards, while continually modifying its own weights to improve the result, a machine learning model can eventually converge upon an optimal solution. That is, it can solve the problem we set it to solve. That's when we can get the model to work, anyway! Building these models can seem more like alchemy than traditional engineering, as it can be a mystifying task to get a deep-learning model to converge. But even when we get the model to produce a solution, there is no way for a human to retrace or mentally represent the solution path. Our minds cannot consciously represent billions or trillions of variables! It is a practical impossibility. This is what is sometimes called *intrinsic opacity* in a "deep" learning model, that is, a model with very many layers and weights. A model like this isn't opaque because we haven't studied it hard enough or long enough. The math is so complex it would remain opaque even if the smartest human on the planet studied it from now until the heat death of the universe.

So, when we gaze in our AI mirrors, we often cannot ask the mirror why it sees what it sees. Even if we design a way to ask, it can be impossible to get an answer that is both reliable (a correct and precise explanation) and interpretable (that is, understandable in human terms). And this is a big problem if what we are using this tool to do is to recommend a lifesaving medical treatment, deny someone a loan or a job, accuse them of fraud, or predict a child's educational achievement—all things that AI mirrors are currently being used to do. How can we trust such decisions if we cannot understand or interrogate them? There are now many agendas in AI research driven by attempts to get around this difficult problem by finding new ways to make model outputs "explainable." But there's a deeper issue, namely, the way that AI mirrors are now limiting our own space for thinking. Unless we address it, we may soon be unable to answer the question I asked at the start of this chapter: What thoughts do the civilized keep?

In 1956, the philosopher Wilfrid Sellars spoke of knowledge taking place in a "logical space of reasons, of justifying and being able to justify what one says."[4] This means that to be knowers—and more broadly to be intelligent in the ways that humans characteristically are—our minds must be more than inert repositories of facts or true statements. Our minds must be active and competent movers within the space of reasons. Being "in" the space of reasons is another metaphor. It just means being in a mental position to hear, identify, offer, and evaluate reasons, typically with others. Just *having* knowledge isn't enough; we must be able to make it, evaluate it, and share it—and reasons help us do this knowledge work. We are addressed in the space of reasons whenever we are asked questions like, "Why have you done this to me?", "Why should we believe that?", "Why should we do that?", "What could justify that?", or "Why is this good?" Try for a moment to envision a society that can no longer meaningfully ask or answer such questions. Can you imagine that? Is what you are imagining still a *society* at all?

To develop even the minimum of shared understanding necessary to answer questions of this kind, we must be skillfully attuned to the standards of evidence and appropriateness that govern the exchange of reasons. We must be cognitively at home within the space of reasons if we hope to understand ourselves and others well; we must find this space familiar and navigable. Usually, humans rely on language to stand in and navigate this space together, although this is not the only way to make reasons understood. Gestures and other signs can convey our reasons too.

As philosophers Bert Heinrichs and Sebastian Knell argued in their 2021 paper "Aliens in the Space of Reasons?", the kinds of computational tools that we call AI are wholly incapable of standing with us in the space of reasons. One of the most powerful yet dangerous aspects of complex machine learning models is that they can derive solutions to knowledge tasks in a manner that entirely bypasses this space. AI tools can produce outputs in the form of reasons (claims, arguments, objections, justifications) that very much mirror our own. But they do this without replicating human thought processes. It's like the difference between building a cabin by cutting down, splitting, and fastening logs, and erecting a prefabricated cabin made of premixed wood laminate that was extruded and baked in a mold. The cabins might look alike if you don't inspect too closely, but if you ask to see the latter's blueprint, there will be nothing to give you. This is why, as we saw in Chapter 1, large language models struggle to defend even the answers they get right. It's hard to rationally defend an answer that wasn't constructed with reasons.

Machine learning outputs result from the model finding the shortest mathematical route from task to solution, which doesn't require following a blueprint of reasons along the way. It's like the difference between a human climbing a mountain by a well-marked and secure route that others can follow, and a robot being flung from base camp to the peak by a powerful catapult.[5] Keep in mind that in this metaphor, the mountain stands for a problem in the real world. While the robot and the human reach the same place, the robot may well arrive faster. That's pretty amazing if what matters is getting there the fastest! But only the human *knows the mountain*, and only the human can reliably chart the route and guide others up its flank.

Philosophers, such as John McDowell in 1996's *Mind and World*, adapted Sellars's metaphor of the space of reasons to the domain of ethical or moral knowledge. So, we can also speak of a space of *moral* reasons. This "space" can be understood in two ways. It's the private mental space where you enjoy the psychological freedom to reflect upon morally salient facts, values, possibilities, principles, consequences, and ideals that might inform and guide your own actions. It's the space you occupy whenever you are faced with a difficult moral dilemma in your life and have to examine the reasons for choosing one action over another. But it can also be a public space where morally salient facts, values, possibilities, principles, consequences, ideals, etc., are entered into a shared moral dialogue. When well used, that space can inform and motivate the decisions of a given community of actors, whether that's two people or two hundred million.

The metaphor of "space" represents an open horizon of moral thinking and choosing. Preserving your access to the space of moral reasons is essential to being "at home" in the moral world, to having morality as a meaningful feature of the life you share with others.[6] Outside of that space, you are restricted as a moral being: you can't readily examine, explain, or morally justify your actions. Some people just avoid entering that space; they choose to live amorally. Others arguably *can't* enter it; this is a plausible interpretation of the pathology of a sociopath. But if none of us could enter or share that space, we could not act as a moral community, and there could be no moral wisdom for us to draw upon together in shaping and steering our societies.

The space of moral reasons is routinely threatened by social forces that make it harder for humans to be "at home" together with moral thinking. Historically, these threats have ranged from authoritarian ideologies that encourage us to leave the task of moral thinking to our gods or our leaders (often presented as a package deal) to cynical philosophies in which morality is portrayed as mere "politics by other means" (or the wasted effort of the hopelessly naive). Today, AI poses a new threat to our moral capacity. Its offer to take the hard work of thinking off our shaky human hands appears deceptively helpful, neutral, and apolitical—even "intelligent." This only makes the offer more dangerous.

Of course, history contains in its pantheon of great souls many voices of resistance to such threats. They include the countless poets, novelists, philosophers, theologians, activists, artists, and protestors who have insisted that humans hold open our private and public spaces of moral reasons. Some remain famous, others were forgotten or never known. Scholars of the European tradition will cite the golden age of Athens and Rome, the Enlightenment era in Europe, and various civil and human rights movements of the twentieth century as times in which the space of moral reasons had to be vigorously defended and held open, often at great human cost. Other traditions and histories tell countless stories of resistance of their own. Today, AI mirrors pose a potentially *global* threat to the space of moral reasons, and we need a new resistance.

The space of moral reasons is shrinking, both personally and in public life, as a result of our growing reliance on increasingly opaque and automated machine decisions. Such reliance itself is not new. Algorithmic automation has been used for decades for everything from credit scoring and loan underwriting, to missile targeting and military logistics planning, to social care and benefits provision. Such decisions have always been exceedingly difficult to drag into the space of moral reasons. Endless mechanisms of human bureaucracy have been invented to keep the reasons for such decisions from being examined or challenged by those affected by them.

But much of modern public and civil rights law is about demanding just such transparency and contestability; over the past century, hard-won gains were made in many countries in key areas of civic life, from lending and employment to criminal justice, voting, and education. Consider as just one example the 1968 Fair Housing Act in the United States, which created routes of redress for many victims of unfair housing discrimination. In their current form, today's AI mirrors gravely endanger this progress. The opacity of automated systems for approving housing or mortgage applications makes the elements of discrimination nearly impossible for an individual to prove, and in 2019 the Trump administration moved to shield users of automated housing algorithms from legal claims of discrimination, so long as a third-party vendor had assured them the algorithm was "fair." In 2023, the Biden administration reversed the move, but it remains unclear how claimants can know or prove they have been victims of algorithmic housing discrimination. This problem generalizes to countless other areas of algorithmic decision-making, from bail recommendations to health insurance coverage.

The danger is amplified by two common elements of automated decisions issued by today's most powerful AI mirrors. The first is their opacity to human inspection, validation, and explanation—often even by experts. The second is the growing ability of generative AI models to manufacture plausible facsimiles of moral reasoning that can lend a false veneer of public legitimacy to AI outputs. What do we stand to lose? We risk losing the space to jointly reflect upon our societies' actions and their moral status. We risk losing the space to publicly contest the rightness, goodness, or appropriateness of high-stakes decisions and policies that have been automated or opaquely steered by AI tools. We risk losing the space to assign moral responsibility to ourselves and others for such decisions and their consequences. Finally, we risk losing the space for moral imagination in exploring new and better possibilities for collective action and social policy.

If we do nothing to prevent them, such losses will come precisely at the moment when converging environmental, economic, and political crises demand collective wisdom and action to preserve long-term human and planetary flourishing. But even in the short term, we can't afford these losses. They risk compromising the still-fragile structures of democracy and rule of law that it has taken centuries to erect. Fortunately, there is nothing inevitable about these losses, and the outcome remains entirely in human hands. AI is currently being used by powerful human actors as a tool to wall off the space of moral reasons. But public movements of resistance can demand that AI be used differently. As we'll see, AI might even be used to open up and grant wider access to the shared space of moral reasons, if we choose. If we don't, we'll increasingly be excluded from that space ourselves. And we must be at home there if we are to fulfill the vital human task of autofabrication: choosing what we shall become next, and how we will solve the political, economic, and environmental collective action problems that today imperil human civilization. These are our cavalry charges; these are the thoughts that we must keep.

The acute danger that algorithmic opacity and automation poses to the space of reasons was foreseen long before the age of deep learning. In his 1955 short story "Franchise," the science fiction writer Isaac Asimov introduced us to Norman Muller. He is an office drone with a "clerky soul" who, in the story's imagining of the year 2008, is selected by Multivac, the artificially intelligent arbiter of American democracy, to represent the electorate in choosing the next president of the United States. By means of an impressive body of calculations opaque even to Multivac's human handlers, the supercomputer Multivac determines that in 2008, it is the very ordinary mind of Mr. Norman Muller of Bloomington, Indiana that can provide the best window into the collective will of the electorate. As Multivac's system administrator John Paulson perfunctorily explains, by interviewing Muller, Multivac will be able to declare with great mathematical precision the winning presidential candidate, just as Multivac declares with unassailable predictive accuracy the result of all elections, "national, state and local." And of course, given Multivac's predictive power, "elections aren't the only things it's used for."

Yet Muller's role in the election is not to express to Multivac his own personal judgment of who ought to be president. This would not be democracy after all, but simply the rule of one. Rather, by posing to Muller a series of seemingly arbitrary questions about matters as banal as the price of eggs, while monitoring Muller's biometric data alongside his answers, Multivac calculates "certain imponderable attitudes of the mind" characteristic of the American voter at that precise historical moment. Knowledge of these "imponderable attitudes," when combined with the trillions of other pieces of data known to Multivac, allow it to compute with great accuracy what the total vote count of the American electorate would be.

The story gives the reader no cause to question Multivac's predictive powers, or its security from tampering or corruption. Still, the system administrator repeatedly emphasizes to Muller their shared civic duty to maintain secrecy about the details of the process, so that the workings (especially the human parts) are insulated from "outside pressures." Like today's AI mirrors, Multivac is an algorithmic colossus that depends for its existence upon countless human technicians and administrators (and human inputs like Norman Muller); yet it is massively opaque. This opacity is reinforced on multiple levels.

First, the computational power and speed of Multivac exceed the grasp of human thought by magnitudes. Second, Multivac's system architecture and internal logic are not isomorphic to human reasoning. After all, by what inferences could you discover the "imponderable attitudes" and minutia of someone's political mindset by asking them what they think about the price of eggs, or whether they "favor central incinerators"? Third, the scales of judgment are incomparable; Multivac performs a statistical analysis of correlations within a massive pool of data about virtually all known facts, whereas human voters are, by nature of our cognitive limitations, far narrower and more focused in our knowledge and reasoning. All three of these features of the fictional Multivac are also found in today's real-world AI mirrors.

Those aren't the only parallels. As with the tightly guarded corporate secrets about the training data, algorithmic techniques, and final weights of the massive AI models owned by tech behemoths like OpenAI, Microsoft, Google DeepMind, and Meta, in "Franchise" the inner workings of Multivac are largely obscured from public view. In the story, American voters know that Multivac calculates the election results on the basis of an interview with a single representative individual, but the details of how Multivac conducts the interview and how the results are calculated (to the limited extent that these are understood by the human system architects) remain tightly kept secrets of the national security infrastructure.

The fictional Multivac accurately mirrors the final verdict of the American voting public, while today's AI tools are being designed to mirror the verdicts of virtually every other kind of human decision process, from hiring and news reporting to the practice of law. Today those outputs are still riddled with errors and weaknesses; but tech companies are investing heavily in piecework labor from human raters, labelers, and trainers in Nepal, Kenya, or Malaysia who might work for a few dollars a day to help the models bring their mirror reflections more in line with our expectations.[7] Those unseen, uncredited workers, the invisible intelligences behind AI, toil not only to steer the models toward more accurate outputs, but also to cleanse them of our horrors: the child pornography, the filmed suicides, the animal torture—work that leaves many scarred for life.[8] Yet they are erased by the same mirrors whose surfaces they polish. Most don't get to see their own lives and languages reflected in the AI mirror. They only get to clean it. They are exposed daily to life-destroying psychological horrors simply to sanitize the reflection of ourselves that the rest of us see. However, no matter how closely they meet our expectations, the reasons behind AI's mirror images remain inaccessible to our inspection. Or, more accurately, there aren't any reasons there to see.

Many years have passed since 2008, the year by which Asimov envisioned us achieving the total automation of the American political franchise. Today, American elections are still carried out by individual voters, notwithstanding the growing interference of bots, trolls, and disinformation armies cranking out political fiction and AI deepfakes for consumption on social media. Yet each of the forms of opacity that Asimov imagined in "Franchise" already exists in the AI-driven decision systems in wide use today—most critically those computational systems dependent upon unsupervised or self-supervised deep-learning algorithms, which pose special difficulties for reliable human interpretation, validation, and auditing.

These systems today are used to identify terrorist threats and targets in voice, image, email, social media and SMS data; to assign risk scores to defendants in bail, sentencing, and parole evaluations; to determine where and when law enforcement personnel are most likely to encounter certain crimes; or to diagnose cancers and recommend personalized treatment plans. Other systems calculate how likely you are to "fit" into the corporate culture and remain with the company to which you have applied, how close a "match" a stranger is to your romantic preferences, how likely you are to repay the business loan you are seeking, or the chances that your kid will thrive at a selective private school. These are the sort of decisions that govern how well or how poorly our lives go. However, in none of these systems can the average user—or, in some cases, even the system regulators, programmers, and administrators—grasp precisely how the decision process is being carried out, or what salient factors are driving the algorithms' results.

In 2018, a poster on Twitter noted that his dad was using Gmail's autocomplete tool as a source of advice by emailing a decision to himself and then checking if the autocomplete gave an endorsement as a reply: "If it says 'That's a plan!' you know you're onto something."

Today, we have AI mirrors in the form of large language models that can not only endorse your submitted plan, but even confabulate an essay giving you plausible "reasons" for approving it. Yet this is just a statistical guess at the kind of approving things a human might say. Trying to find the "thoughts" behind an AI verdict is like looking for the body on the other side of your bathroom mirror.

In the Asimov story, we are led to question how the political franchise of the voter can ever be preserved under such conditions of algorithmic opacity. But we must notice that one of the most disturbing aspects of the scenario—and our present reality—is the contraction of the space of reasons that it fosters. Norman Muller has no cause to explicitly think through his own political judgments. First, because he assumes he will never be the one American chosen to be directly consulted by Multivac, and second, because even when he is in fact the one chosen, Multivac does not need to explicitly ask him for his personal opinions about politics or the good of the union—much less ask him to justify those opinions with reasons.

Public political dialogue is similarly superfluous in "Franchise." It still happens, of course, insofar as the story tells us that politicians still campaign and voters still form opinions, but the causal link between explicit public reasoning for those opinions and the final vote is lost. Why would voters feel any need to understand their neighbors' reasons—or explain the reasons behind their own opinions—when Multivac can correctly predict everyone's votes by entirely indirect and opaque means? In the story, the background assumption is that Multivac's predictions are, if not perfect, at least as accurate as the tallying of millions of actual human votes—and far less costly and cumbersome.

The same justification is given for the use of AI mirrors to automate life-and-death decisions in human institutions today. No credible AI researchers are claiming that any computer currently in operation can actually grasp the moral, legal, or political gravity of drone targeting decisions or sentencing recommendations, much less reason wisely about them. But if an AI tool can make decisions—only faster and more cheaply—that are just as reliable as those made by the flawed humans who *do* reason about them, then the logic of efficiency invites us to let the reasoning drop out of the process. It now looks like an unnecessary human excrescence of analog decision-making. It doesn't help that the present quality of public reasoning and decision-making, on social media and in our political congresses and parliaments, invites the consolation of cynical despair. Maybe humans are just not cut out for good reasoning. At this point, how much worse could the machines really do?

What do we lose by giving in to that logic? We have already mentioned the risks to our capacities for shared moral reflection, collective moral appeals, jointly allocating responsibility, and exercising moral imagination about our futures. Remember also that virtues like moral imagination and responsibility have to be cultivated through practice; we acquire virtues by doing, not by mere wishing. If automated reasoning crowds out our own opportunities to practice moral and political reasoning, it leads to what I have called "moral deskilling"—the moral equivalent of a world where GPS means that no one knows how to find their way with a map.[9] As long as the satellites keep working, we might be willing to accept that particular trade. Alfred North Whitehead might also have taken that bargain as the price of progress; navigation is now one less thing we have to do. But should we accept the same bargain to hand over our capacity for moral and political deliberation? What more of us is left?

To a virtue ethicist like me, it is obvious that preserving ample space for moral reasons, both private and public, is an essential prerequisite for the acquisition of the virtue of practical wisdom. That's the virtue we encountered in Chapter 3, the one Aristotle called *phrónēsis* and that we used to translate into English as "prudence." This virtue, which matures and harmonizes all the rest, enables excellence in both private and public reasoning about moral and political matters. But I can't develop it, and you can't either, if we are denied the opportunity to engage in such reasoning ourselves and learn from its consequences. On top of that, practical wisdom is the virtue that allows for moral and political innovation. Because it links our reasoning from prior experience to other virtues like moral imagination, it allows us to solve moral problems we haven't encountered before. We have a lot of those coming at us today.

As the philosopher Hans Jonas wrote in 1984's *The Imperative of Responsibility*, until the twentieth century, humans never had to confront a moral question like, "How do we balance the needs and interests of living human individuals with the future of life on the planet?" The fate of the latter was simply not up to us—not until our nuclear weapons, bioengineering, and industrial fossil fuel extraction put it in our power to snuff out the planet's living light. But now that we have that power, we have the responsibility to decide how to use it. And that task is going to require every ounce of practical wisdom we can collectively muster.

Practical wisdom or phrónēsis in the light of this responsibility is more than just calculative reasoning about how to optimize our personal and collective survival. It is also existential reasoning as autofabrication—deciding what kinds of beings and communities we ought to work to become. More than one version of human power could manage to survive our peril. One version might survive through cruelty and exploitation of the most vulnerable. Another might survive through expanded solidarity and moral courage. What versions of ourselves, from all of our many possibilities, will we choose to make real? This is the task that wisdom must guide.

But as Hans Jonas saw, "we need wisdom most when we believe in it least."[10] Today, we are shrinking the spaces for moral wisdom to be cultivated and exercised, at exactly the moment when the future of the human family depends on us having it.

Algorithmic decision systems have long presented this risk, but today's more powerful AI mirrors amplify it. Unlike earlier modes of algorithmic decision-making that were opaque to decision subjects but well understood by the human bureaucrats that designed them, the precise causal chains that produce today's deep-learning decisions can be inscrutable even to the most talented AI researcher. It's hard to demand an explanation from someone in power if they don't have it and couldn't get it even if they wanted to give it to you. For that reason, opaque AI decision systems are highly attractive tools for those in power; they offer a virtually bulletproof accountability shield.

There is one final way that today's AI mirrors amplify the danger of algorithmic decision systems. It's because, unlike older algorithmic tools—which give a result with a label or confidence value but are otherwise mute instruments—today's AI mirrors can and do *bullshit*. In his 2005 book *On Bullshit,* Harry Frankfurt diagnosed bullshitting as among the most pernicious of contemporary threats to truth. The book became one of the few contemporary philosophical works to immediately resonate with popular audiences, and its message is no less timely now.

The core of Frankfurt's argument was a distinction between a lie and bullshit. When a liar tells a lie, they are fervently hoping you will believe the lie rather than the truth. The intent to deceive, then, gives the liar a strong interest in the truth. They have to keep very careful track of the difference between truth and falsehood, in order to make sure you don't get near the former. Truth therefore matters to the liar. They understand its scientific, moral, and political value, which is why they try to keep it out of your hands. In contrast, Frankfurt observes, the bullshitter simply doesn't care. That is, the bullshitter is indifferent to the distinction between truth and falsehood. A bullshitter doesn't even stand with us in the space where truth matters. They're just talking to hear themselves talk, or to appear knowledgeable, or to remain in a position of power.

Frankfurt's point is that bullshit is even more dangerous to our social foundations than lying. The practice of lying is usually selective and carefully crafted. In the hands of all but the compulsive liar, who will quickly be recognized in society as untrustworthy, lying is a sharp tool. If you want it to serve its purpose, you can't use it all the time. You have to keep track of which lies you tell, and to whom. Bullshit, on the other hand, tends to be spewed indiscriminately—and for some, it's a 24/7 operation. It's the favorite tool of modern media manipulators who take their lead from Steve Bannon, former head of right-wing media outlet Breitbart news and chief strategist for Donald Trump. In 2018, Bannon famously confessed to adopting the strategy of "flooding the zone with shit." To keep the media off the scent of a story, you don't bother to craft a careful lie that needs to be protected. You just drown the public conversation with massive quantities of bullshit, so that no one can even find the story—and if they do, they can't tell the difference between it and fiction. Flood the zone often enough, and people will stop even trying.

What should worry us most about today's AI mirrors, particularly generative large language models, is that they are hard-wired for bullshit. That is, they are not like traditional search algorithms, built to ferret out useful facts and serve them to you. Nor are they designed to lie to you. They are simply built to generate fact-like patterns of language—plausible statements that sound like what a person might say to a given prompt. Whether the outputs are true or false is of zero importance for a generative model. In this sense, they are like the human bullshitter. They aren't designed to be accurate—they are designed to *sound* accurate. Many worry that this makes them perfect tools for "flooding the zone" with even more bullshit than our media environment already has floating in it. As we saw in Chapter 2 examples of AI fabrication, the problem is real: autogenerated ChatGPT bullshit can now be found in legal filings, news articles, social media posts, scientific preprints, and countless plagiarized student essays.

This gravely endangers the space of moral reasons and our opportunities to inhabit that space with others. One obvious reason is that even if I'm thinking carefully with other people about matters of public concern, many of their views, or mine, might be AI-generated bullshit we picked up—and mistook for truth—in something we read or heard. Another concern is that the flood of bullshit generated by these tools, on top of the social media tidal wave of human-generated bullshit, may just render most of us helpless to sort through it, leading us to stop even trying to base our decisions, political positions, and public policies on facts. At that point, whoever already has the most power, influence, or resources wins the day; there's nothing to discuss.

Worse, these new AI mirrors can be fine-tuned in such a way that they succeed in passing off their bullshit as legitimate reasoning and argument, producing mindless sophistry that would have made Socrates's foes weep with envy. In that scenario, algorithmic bureaucracies that today remain opaque and unanswerable to their citizens suddenly start "talking"—through the mouthpiece of AI mirrors that have perfected the job of telling citizens whatever they want to hear. Fortunately, the kind of bullshit that our AI mirrors manufacture today tends to be detectable from its generic quality, not unlike the car salesman whose smooth, oily pitch has all the spark of mayonnaise. As Jessica Riskin said in *The New York Review of Books* of the AI-generated student essays she was marking, the flat, entirely plausible, and competent words of an AI mirror often read like the "literary equivalent of fluorescent lighting."[11]

But this may change even by the time you read this, as model designers begin to leverage the idiosyncratic tastes of intelligent human raters to fine-tune these models until they mimic our thinking with more human spice and character. We may then more easily find ourselves talking mostly to our AI mirrors—not realizing we have been led out of the one space in which our moral and political capabilities can grow.

It's worth backing up, however, to see how much we've already allowed algorithmic automation of moral decision-making—using much simpler versions of AI mirrors—to diminish our shared humanity. A prime example of this is the use of proprietary algorithms in judicial decisions. As revealed in the landmark 2016 investigative series "Machine Bias" by *ProPublica,* many states in the US employ risk algorithms for bail granting and other judicial decisions.[12] Many of these are, in principle, interpretable models—that is, they are not intrinsically opaque in the way we described earlier in this chapter. Yet in most cases, risk scores for individual defendants are given directly to judges and parole boards with no transparent analysis of their basis or their limitations, and no required training or standards for responsible and effective use. Neither judges, nor defendants, nor reporters can typically gain access to the algorithms themselves.

Instead, the *ProPublica* team conducted their own analysis of the output of Northpointe's COMPAS algorithm for predicting recidivism risk in criminal defendants. They found that it falsely predicted Black defendants as reoffenders at almost *twice* the false-positive rate for white defendants. Similar racial disparities have been found in many other algorithmic systems all over the world. Remember:

using AI mirrors to automate an already unjust system will not only reflect the existing injustice—it will often magnify the injustice and then project it across every decision.

The COMPAS survey instrument does not ask about the defendant's racial background, so the bias comes into the analysis via unknown proxies for race. We saw this phenomenon in Chapter 2; AI mirrors don't need our biases clearly labeled to find and reproduce them. Further scholarly analysis of racial bias in tools like the COMPAS algorithm suggested an inevitable design tradeoff in these kinds of algorithms between racial parity in false positives and parity in true positives.[13] That is, in many cases you can have racial equity in terms of the algorithm's accurate predictions of high risk, or you can have racial equity in terms of who the algorithm's errors hurt, but you can't have both.

If that's right, then here is an opportunity for substantive moral and legal deliberation: about the value of due process, the presumption of innocence, and the particular social harms and costs of false positives that tilt toward defendants of color. But this conversation could not happen because Northpointe refused to share the specifics of its proprietary algorithm. How then could researchers' suspicions about its design limitations be confirmed? As a result, the moral and legal discussion we desperately need to have about these tools remains stalled. This also hampers the ability of any defendant to present a reasoned argument that the tool's use in their particular case introduced bias; it likewise denies critics the leverage to publicly reason with the company or lawmakers about an appropriate remedy. By closing off our access to the space of moral reasons, these tools block our ability to appeal or contest the rightness, goodness, or appropriateness of algorithmically mediated judgments, which is of course why unjust institutions seeking to evade criticism have such a strong interest in keeping us out of that space.

The space of moral reasons also allows humans to be responsive to new moral reasons communicated by others, or to other new moral information in our environment. As long as the process of thinking and choosing remains open, there is time for humble correction—or for deepened resolve. Even once I decide, the wisdom of a new voice or insight can move me to revisit, retrace, then modify or retract my decision, just as I can retrace the steps of a hike that I made yesterday and take an improvised detour. Even when the act lies in the past, or was done without thinking, I can use the space of moral reasons with remedial intent: to give a quick, emotional decision a careful moral audit when time allows, allowing me to understand how my habits and raw impulses are serving me and others—well or poorly.

Consider the Multivac scenario, in which a precise and unambiguous verdict predicts the election outcome without any need for the messy vote itself. It's a bureaucrat's dream. Now contrast that story with American presidential elections. The moderate transparency of American voting patterns (through turnout data, exit polls, local vote totals, voter interviews, and other indicators) enables public reasoning—however chaotic and discordant—about which social factors, groups, and events were responsible for a result. In the United States, the power of evangelical Christian churches, the Black Lives Matter movement, eroded voting and abortion rights, disaffected millennials, the Trump legacy, the disenfranchised working class, white supremacists, and QAnon conspiracies on social media may not produce anything like social cohesion or reconciliation. But our clumsy analog voting practices, unlike the fictional Multivac's, do still afford space for public and private reasoning about the causes and merits of our moral and political choices. In fact, the chaos evident in many of today's election discourses reveals important truths about just how deeply the world's formerly dominant political and moral visions have splintered. Even looking at a mess is more instructive than staring into a black box!

No one is about to launch a Multivac today. But AI mirrors that are no less opaque than Multivac are increasingly used for large-scale corporate and institutional decision-making, such as the sort of hiring software now used by human resources departments in most large organizations. A 2016 *Harvard Business Review* article estimated that up to "72% of resumes are weeded out by an algorithm before a human ever sees them."[14] In 2018, 67 percent of hiring managers and recruiters surveyed on LinkedIn said they were using AI to filter out job applicants and save time. Those figures are now far higher. In theory, hiring algorithms are supposed to promote a more diverse and well-qualified workforce, bypassing irrelevant factors that human evaluators commonly favor or disfavor but that do not reliably correlate with candidate quality, such as European or male-sounding names. In practice, however, hiring algorithms often reflect, perpetuate, and even magnify these same biases embedded in their training data.

A 2021 Harvard Business School report found that automated résumé screening and candidate evaluation systems were contributing to a "broken" labor market, with millions of well-qualified applicants being screened out and discarded by algorithmic filtering that is often corrupted by historical bias.[15] If, for example, a machine learning algorithm is trained on data about previous workers in a given industry and "learns" that male engineers in the training data set were recruited more often and promoted more quickly, the algorithm is likely to unfairly favor male candidates for engineering jobs going forward. Unless specifically programmed to avoid that trap (easier said than done in machine learning), the AI mirror will reproduce the historical pattern.

The AI mirror cannot know, and cannot learn, that data is likely to reflect historically ingrained but unjust social biases against women engineers. This is because it doesn't understand history—or bias, or justice. The system simply cannot tell the difference between a good reason for a person being hired and promoted, and a bad (but widely used) reason. That's because our AI mirrors aren't standing in the space of reasons with us at all. This was how Amazon ended up in 2017 with a recruiting algorithm it had to scrap because it had learned to downrank applicants who had attended women's colleges, or who had served as the president of a women's chess club. The same can happen for past hiring biases based on economic class, region of origin, or prestige of university background. It's why some applicants now hide words like "Harvard" or "Cambridge" in a résumé's margins in white font, in the hopes of gaming an algorithm's prejudicial inclinations toward Ivy League applicants.[16]

One popular service, HireVue, uses opaque proprietary algorithms to analyze video interviews, game-based assessments, and chatbot conversations with job candidates, and then predict traits like "adaptability" and "leadership potential" with the company. If a candidate is rejected by the system, how confident can we be that this is related to a job-relevant personality trait, such as quickness to anger or deceptiveness, as opposed to the algorithm's marking of an unusual muscle tic, a regional accent, culture-specific facial expressions or gestures, body mass, external signs of age or disability, or race-related dialect? All of these are unethical and/or illegal reasons to discriminate against an otherwise well-qualified job candidate.

Yet while HireVue boasts that their technology actually mitigates human bias, there is no way of knowing whether HireVue's algorithm discriminates on these bases. You might point out that human interviewers routinely discriminate on such bases. This is true!

But at least we do not assume human interviewers to be objective analysts, and we are able to ask them about their reasons. A human hiring committee can hold a member's feet to the fire to explain the basis of their reflexive dislike or distrust of a candidate that others on the committee find highly qualified. And the committee can discount that member's judgment if they repeatedly fail to give good reasons supporting their vote. No such exchange in the space of reasons can take place between a human hiring manager and the HireVue algorithm.

Imagine that Ophelia has applied to a large engineering firm that uses such a hiring system to automate the screening and interviewing of job candidates. Now say that Ophelia's application was rejected, despite the fact that a male friend with far less experience and weaker credentials got an interview. Who or what is responsible for the decision? How can she know? Who can she ask? The human resources representative? In a previous era, Ophelia would have known that this is the person who owes her an answer. The HR representative might lie about their reasons, but regardless, there is a person answerable for the decision, who *can* give reasons. Yet in the automated hiring chain, who gives answers to Ophelia? No one can.

The company's employees will not be able to reproduce the specific calculations that caused the hiring algorithm to exclude Ophelia. The algorithm itself will be proprietary, and if the model was trained using deep learning and other opaque AI techniques, even the software engineers and data scientists who created it will not know exactly how or why it works in a given case. So, perhaps Ophelia was rejected because the algorithm is unfairly biased against women (in which case those who designed and trained the algorithm would be to blame). Or perhaps she was rejected because some combination of her answers was strongly correlated in the training data with cases of employee theft or embezzlement. Or perhaps Ophelia went to the University of Edinburgh and there is a freak statistical cluster of Edinburgh grads in the data set who washed out of the job in their first year. We can imagine a thousand other possibilities. Some will track good reasons not to hire someone, and some will track bad reasons. If Ophelia doesn't know which applies to her, and no other humans know either, then we have all lost the space to reason about, and hold the hiring decision-system accountable for, its impact.

Because the space of moral reasons is essential for reflecting upon, contesting, and holding ourselves and others accountable for personal and social choices, we need to be sure that our AI mirrors are not used to refuse us entry to it. Yet the space of moral reasons also has a final, forward-looking function. It enables and encourages the use of moral imagination in considering and weighing alternative patterns of moral reasoning and judgment. To reason about my past moral choices is always to invite the moral counterfactual: "What if I had done/chosen/said *that* instead?" Often this begins in the space of reason-giving and reason-commanding that take place between self and others, or the self and its conscience, or the public and the public conscience: "Why didn't you/we do *this*?", "Why don't you (or we) want/value *that*?"

To answer such questions, we often have to construct an alternative history in which different motivations and thoughts lead to a different decision. I (or we, in the case of public reasoning) may decide that these alternate motivations and alternate reasons are ultimately wrong or incoherent. I may conclude that I (or we) still would not do that other thing. But moral learning and growth takes root in the space of moral imagination where I, or we, can realize that other, better, choices were available to us, had we exercised other and better patterns of reasoning, feeling, and valuing. Here we may resolve next time to reason better and *do* better—to give more discerning sentences, hire more fairly, vote more responsibly, live more sustainably, fight more bravely, or love more fully. This is the fertile soil in which futures are grown.

Yet when AI mirrors crowd us out from the spaces in which we chaotically, inefficiently, and very imperfectly think about what we do—when we tolerate living as passive machine inputs and beneficiaries of machine outputs—that fertile soil is washed away. With it are lost our possibilities for autofabrication—through meaningful moral reflection, moral appeals, moral responsibility, and imagination. What are we without these? Only what we have already been. In the words of the twentieth-century computing pioneer Joseph Weizenbaum, "a computing system that permits the asking of only certain kinds of questions, that accepts only certain kinds of 'data,' and that cannot even in principle be understood by those who rely on it, such a computing system has effectively closed many doors that were open before it was installed."[17]

While it is typically impossible to design predictive algorithms and other AI decision tools to be optimally fair and accurate across all criteria and contexts, many researchers are discovering ways to make AI systems more accurate, reliable, and fair. These include building models with more diverse teams and stakeholders, anticipating likely biases and social effects that may need to be mitigated, auditing outputs for disparate or unjust impacts on marginalized and vulnerable groups, and "shifting the burden of uncertainty" from impacted groups to decision-makers, in order to incentivize AI designers and users to seek out better and more relevant data with which to train the algorithms.[18]

As useful as such recommendations may be, they evade the concerns raised in this chapter. For even if designers and users can be incentivized to promote better, fairer social outcomes in the use of AI, this might still be done without any concerted effort to make the systems themselves more transparent to users or publics, or to keep human reasoners engaged in decision processes of moral and political gravity. Moreover, it doesn't stop us from flattening our own decision-making to the dimensions projected in the AI mirror. Henrik Skaug Sætra tells us that machine mirrors reflect an ideal in which "the human realities of emotions and their connection to human decision-making is stowed away, neglected, and at times forgotten."[19] Yet as he notes, decisions made without emotion are often much poorer for it. Ethical design and use of AI mirrors of human decision-making will therefore require more than fair, equitable, and accurate outcomes. Nor is transparency enough, if entry into the space of moral reasons is blocked by other means. We have to recognize the intrinsic value of human participation in moral and political thought.

In a future where this value is recognized, AI developers tasked with building a decision tool would ask themselves: What kinds of thinking does this system mirror or duplicate? Are those thought processes of no real value? Or are they among the thoughts that we must keep? Regulators of automated decision systems would ask: What justifies the impact of this system on the space of public reasons? Deployers of these systems would ask: How can this tool be used to preserve or augment, rather than shrink or eliminate, the space for human thought and reasoning in this organization? AI researchers would ask: How could the computational power of AI expand the space of reasons for people, and enable wider, more effective and equitable access to it? In a world where "machine thinking" is the benchmark for human beings and computers alike, these questions don't get asked. In a world where the space of reasons thrives, they must be.

While Asimov's Multivac is not in our immediate electoral future, very close cousins of it are already taking shape in many other areas of personal and public decision-making, about morally and politically significant matters of health, justice, labor, finance, education, family, and community life. In more and more of these domains, the personal and public space of moral reasons is contracting, as the power and projected socioeconomic utility of AI expand.

The space of reasons has been constrained before—by priests, kings, oligarchs, and family elders who would gladly substitute their moral and political judgments for ours, and by bureaucrats who endlessly invent analog means of rendering their own judgments opaque. But at the heart of the modern Enlightenment lies the eighteenth-century philosopher Immanuel Kant's urgent call: *sapere aude!* He asked us to dare to think for ourselves, together. It was a call answered in part by the rise of modern public education and liberal democracies that, for a time, promised to expand the space of moral reasons and its privileges to the greater share of humanity. Today we risk losing that inheritance to the algorithms engraved in our "helpful" AI mirrors, which we are told can progress civilization by performing all the critical operations of moral and political life without having to think about them. Unlike the tyrants defeated by past movements of democratic resistance, these tools are presented to us by their corporate makers not as selfish oppressors, but as benign and neutral agents of our will. This only makes them more dangerous.

If AI mirrors make human operations in the space of reasons seem too inefficient, unreliable, and unnecessary to be tolerated, then AI's impact on the moral and political maturity of humanity may be more damaging than that of the worst dictator. Fortunately, this future is not set for us. Those of us who would fight to protect the space of reasons—so that we may together engineer our own future selves and communities, and keep laboring to craft better worlds than those that have passed—have a long history of resistance to learn from. Those who came before would tell us that the prize has always been worth fighting for.

Chapter 5

The Empathy Box

Dost thou know why the mirror (of thy soul) reflects nothing?
Because the rust is not cleared from its face.

—Rumi, "Masnavi" 1:34 (1273 CE)

All technologies are mirrors, because all technologies are extensions of human values into the built world. Every technology—from the wheel, to the book, to the engine, to the computer—reflects what humans of certain times and places thought was worth doing, making, enabling, improving, or trying. Judgments of "worth" always indicate a human act of valuing, whether or not we are valuing the right things. Even acts of violence can express a value judgment. If I willingly strike you in the face, I am expressing a judgment that at that moment, your personal dignity and well-being are not worth enough to me to command my respect—not valuable enough to warrant me making the necessary moral effort to restrain my rage.

When we gaze in our AI mirrors at the words and images they generate, at the predictions and classifications they make, we are not seeing objective truths, under any definition of "objective." We are seeing reflections of what humans valued enough to describe or record in data. But not *all* humans. What our AI mirrors show are the values of those humans who have historically had the power to shape the dominant patterns now engraved in our recorded data. For example, for most of recorded history, women were largely barred from professional roles that involved authoring books and magazine essays, producing scientific research, writing and directing films, or reporting news stories. While some countries have now opened these roles to women, any large language model trained on a corpus of digitized text that goes back more than a few decades will reflect primarily what *men* have had to say, and what men have valued enough to bother to describe. The same is true of the corpus of visual art, music, and other domains of culture. And again, these cultural reflections certainly won't reflect the value judgments of *all* men, or even most. Most men who have lived on this planet have been economically, racially, or otherwise marginalized in ways that also blocked their contributions to today's digitized cultural record of the human family.

The point is that our AI mirrors are nothing like neutral reflections of a shared human reality. They are very potent indicators of how a small subset of humans have seen and valued the world, and the marks they have left on it. For many reading this book, the reflections in AI mirrors will resemble something not too far from your own assumptions and value judgments about the world; what you see in them will be mostly comforting and familiar, even as they surprise and delight you with their power to speak without faces, or to write sonnets without hands. For most others, these reflections have a dimmer cast. Their voices speak a language that is not originally yours, or their patterns retrace a historical arc of devaluation and denigration by other humans who have chosen to see you and your kind as lesser.

This character of AI mirrors can be immensely instructive: a powerful source of learning. Before you can change the arc of history, you have to study its pattern. Like Rumi's mirror of the soul in his poem "Masnavi," the surfaces of our AI mirrors are deeply rusted by our own histories of injustice and oppression. If we don't see in them what we wish to see in ourselves, we must clear that rust away. The reflections in AI mirrors can therefore be a powerful driver of the changes needed to bring the human family back from the brink of our self-caused economic, geopolitical, and environmental crises. They can reveal subtle patterns that humans cannot otherwise see—or they can make undeniable the patterns we've been too ashamed or too comfortable to acknowledge.

For example, AI mirrors can reveal deep and pervasive patterns of unjust discrimination in systems of so-called justice, by providing incontrovertible evidence of institutional racism in policing and courts—in the very countries that applaud themselves as global champions of human and civil rights. AI mirrors trained on environmental data can link patterns of accelerating crop failure to climate change, and tie growing infant mortality in vulnerable communities to the behavior of agricultural polluters and mining operators. They can surface patterns of white-collar crime and political corruption in a vast sea of global financial transactions.

One powerful example of this comes from the health domain. We saw in Chapter 2 the racially biased hospital algorithm used in the United States that steered critical care away from Black patients, and toward white patients who were at less risk. But AI mirrors can also be used to *detect* medical racism. Of course, we could just believe the clear testimony of the communities that experience it every day, and then get to work fixing it, which would be a lot cheaper, fairer, and quicker. But it is precisely medical racism which ensures that the lived experiences and reports of medical racism by patients are routinely dismissed or ignored by medical professionals. It's a vicious cycle. AI mirrors can be one tool in our efforts to break that cycle by providing the kinds of evidence that medical professionals are already trained to accept.

How might that happen? A 2021 study on knee pain in the journal *Nature Medicine*, led by two of the researchers behind the 2019 study on the hospital algorithm, looked at a long-unexplained "pain gap" affecting Black and other underserved patients in the United States.[1] Black patients tend to report more knee pain than white patients, even adjusting for the severity of knee condition as rated from x-rays. The study showed that a machine learning algorithm could predict the greater pain experienced by Black patients simply by looking at radiographic images of their knees, proving that the AI tool could see physical signs of injury that human radiographers were missing. That is, the study provided evidence that the causes of the greater pain were in the knee, not in Black patients' imaginations or manner of describing their pain or any other cause. It put the responsibility—for closing the pain gap, which results in denial by insurers and clinicians of vital treatment for Black patients—back where it belongs: with medical professionals who weren't listening to what Black patients were telling them, instead overly relying on imaging diagnostic techniques that were developed and validated decades ago on an unrepresentative population of primarily white British patients. Our AI mirrors can replicate our unjust biases, but they can also expose them to light and boost their victims' legitimate demands for restitution.

However, these uses of AI are far too few and underfunded. Today's AI mirrors, particularly those developed for commercial purposes, are most often carelessly designed and used in ways that actually reinforce and deepen the dominant historical patterns of human valuing and acting that we already know to be unjust, unsustainable, and corrosive to our societies. The magnifying power of AI mirrors means that they have to be used with the expectation that they will amplify harmful patterns unless deliberately made to do otherwise. Because what will you get if you naively train an AI model on an unjust medical or financial or criminal justice system? An AI tool that calculates how to be even more efficient than people at delivering injustice.

Countless other examples of algorithmic bias have manifested in AI mirrors, as we've already seen in Chapter 2. But it's not only the newest and biggest AI tools that cause these harms. A simpler algorithmic mirror in the Netherlands forced the Dutch government to resign in 2021 when it was revealed that tens of thousands of families—disproportionately immigrants—had been unjustly accused by the algorithm of child benefits fraud. In that case—and in cases of similarly flawed fraud detection algorithms in the US, UK, and Australia—the high price that states and vendors paid in lawsuits and bad press was predictably dwarfed by the unfathomable price paid by the innocent: suicides, job losses, homelessness, and families ripped apart. In the Netherlands case, at least 1,675 children were taken from their parents, solely on the basis of a demonstrably flawed algorithm. Hundreds have still not been returned.[2]

Like glass mirrors, AI mirrors don't just magnify, they distort. AI mirrors sometimes produce distorted outputs due to random or unexpected algorithmic variations; at other times they do it in response to signals from our own distorted values and ideals. We've seen that the latter can create runaway feedback loops with strange or damaging results. For example, the early images of people produced by generative AI tools like DALL-E and Midjourney featured strange alien-looking women with the same distorted features: impossibly thin frames, sharply jutting cheekbones so high they almost met the eyes, and smiles with a truly unsettling number of giant white teeth. These tools had learned to exaggerate the dominant features of women's images in the data set in a way that yielded an "uncanny valley" of model-esque weirdness. Those distortions in the training data were likely fed by prior rampant use of Instagram beauty filters, the design of which is in turn conditioned by the distorted standards enforced for decades by Western beauty magazines and advertising. Now those distorted images of artificial AI-made women are appearing in magazines like *Vogue Singapore* and are used by advertisers like Levi Strauss & Company in place of professional human models. Like the ancient allegory of the ouroboros, AI closes the loop—a serpent eating its own tail of distorted, unattainable ideals.

Like the distortions produced by glass mirrors, the distortions in our AI mirrors can be accidental defects, but they can also be deliberately engineered. Distortions can be highly useful: think of the side mirror on your car that is designed with a slight convex curve to give the driver a better view of what is approaching from the rear. They can also be very dangerous. If I don't know my mirror distorts my view, I can make a fatal error. And unlike automobile mirrors, today's AI mirror distortions are bending perceptions of our social and political reality, not just our physical surroundings.

Through the algorithmic feedback loops operating in the digital media spaces that have come to function as the new public square, AI mirrors amplify and normalize our biases, reinforce our most polarizing opinions and most aggressive stances, and boost the visibility of our most uninformed "hot takes." In doing so they reflect back to us images of human civic agency so distorted in their form that they not only shift the "Overton windows" of acceptable political conduct but make us lose our already fragile faith in the human capacity for political wisdom. The result is that actual political behavior drifts closer and closer to the originally distorted mean, producing as a self-fulfilling prophecy the very reality that our distorted mirrors predicted.

This is why it is so important—for purposes of both safety and transparency—that we know when our mirrors are distorting reality.

There's a very good reason that the distorting side mirror has a little warning engraved on it: "objects in mirror are closer than they appear." In contrast, AI mirrors today are rarely tested rigorously to find the distortions they produce, and they almost always lack the safety and transparency signposts and guardrails we need. Until we demand these, it will be increasingly difficult to truly know ourselves. And when we can no longer know ourselves, we can no longer govern ourselves. In that moment, we will have surrendered our own agency, our collective human capacity for self-determination. Not because we won't have it—but because we will not see it in the mirror. Later, we'll turn back to this danger, and how we can still avert it. But first we must grasp the other core dimensions of the human personality that may be amplified, distorted, or occluded in the AI mirror.

Glass mirrors have long been a helpful tool of the magician and the charlatan; their power to produce illusion is both entertainment and weapon. Today's AI mirrors are both as well, and their impact on the creative dimensions of the human personality—our capacity for storytelling and meaning-making—is now as profound as their impact on our political agency. While creative work was long thought to be among the human labors safest from AI-driven automation, generative AI tools violently upended that conventional wisdom. Creative professionals remain deeply divided about whether the latest generative AI tools will amplify their talents or devalue and replace them.

AI mirrors can be used to suggest new and more imaginative possibilities for artists to explore. For example, they might offer a dozen surprising plot turns a writer can contemplate for her main character, which she can develop or modify according to her muse. They can also aid inclusivity by lowering barriers to enter creative work. For example, a generative AI application can help people without industry mentors or expensive film degrees turn a novel idea into a professionally formatted screenplay or compelling pitch deck. Yet the same technology can be used by film and television studios to break a writers' strike, filling the void with cheap, auto-generated scripts perfectly fine-tuned for maximum audience engagement. Which path is better favored by the media ecosystem's current economic incentives and values?

This should remind us that AI mirrors don't only reflect the human values embedded in their design, training data, or chosen optimization metrics. Even for AI tools with considerable *polypotency*—the potential to be used in many different ways, by different people, to accomplish many different things—the uses and effects that actually come to pass are typically steered by the dominant values of the social context in which the tool emerges. This is what philosophers of technology have long called the *affordances* of a given artifact. An affordance isn't destiny. Just as a gun doesn't have to be used to commit a violent act, but more likely will if placed in the average American family home, a generative AI model doesn't have to be used to replace human creative or professional talents—but likely will if commercial incentives favor it. In principle, creative AI tools can just as easily enlarge human talents. But in markets where companies are desperate to boost their next quarterly earnings in a race to the bottom for investors' and shareholders' fickle affections, the affordances of AI do not tend to favor the paid professional.

That said, there is a deep disanalogy between what AI mirrors do and what human creatives do, one that we can easily lose sight of when under these same cultural and economic pressures to produce, produce, produce. First, creativity is not just a matter of varying your learned pattern, churning out slightly different versions of what has worked before. But that is exactly what generative AI models are trained to do—and to be honest, it is a fair amount of what creatives do in their daily work. If you're an accomplished painter, writer, musician, dancer, or poet, you likely have a distinctive and recognizable style that your audiences respond to, and that you keep giving them, with just enough novelty threaded in to keep them from getting bored. But eventually the creative who does nothing more than this fades away. For your loyal audience will eventually discern the pattern too, and when they do, they will find even your variations increasingly predictable and dull. The arc from inspired genius to mediocre hack can be painfully short.

Moreover, the heart of creative work is not creation, but expression. To merely create is to bring into being what was not there before. It is now a fairly trivial operation for an AI model to "create" in this limited sense, by producing new variations on an existing data set. To *express* is different. To express is to bring into existence something that speaks of something else. An AI tool can create a new sea shanty or a new sculpture or a new abstract shape. But what can it express through these? To express is to have something inside oneself that needs to come out. It pushes its way out: of your mouth, your diaphragm, your gesture, your rhythmic sway. Or you pull it out—because it resists translation, resists articulation.

A generative AI model has nothing it needs to say, only an instruction to add some statistical noise to bend an existing pattern in a new direction. It has no physical, emotional, or intellectual experience of the world or self to express. As long as we recognize and value the difference between mechanical creation and self-expression, AI poses no threat to human creativity or growth. But there lies the problem, in light of the cultural and economic values that currently shape AI's affordances. Increasingly, as we saw in Chapter 3, we ourselves are measured in terms of our ability to resemble our own mechanical mirrors. Whether it's the pressure to produce your next album, or publish enough to get tenure, or film enough videos to keep your subscribers—our dominant values favor those who don't get writer's block, who don't struggle to find the right words, or images, or notes, or movements, who never get caught up in the swirling drag of inexpressible meanings. Our economic order has long rewarded creators who work like machines. Should we really be surprised that we finally just cut out the middleman and built creative machines?

Remember: the danger to our humanity from AI is not really coming from AI itself. The call is coming from inside the house; AI can devalue our humanity only because we already devalued it ourselves. Indeed, one very common response to the cultural shock of generative AI tools has been not to admit the discontinuity between humans and predictive machines, nor even to pretend that AI tools have their own inner unspoken depths, but rather to deny that humans do. When GPT-4 was released, I became involved in a revealing exchange on Twitter. The original poster stressed the wide gap between human experience and GPT-4, a tool that is merely a mechanical generator of "very good predictive text." One poster replied, "so are we." In fact, many replies challenged the original claim of difference. I tweeted about my dismay at "how many humans already believe they are merely a machine for generating valuable word tokens." While many replied in sympathy, others insisted that humans are "primed automatons" with mere delusions of holding greater depths.

Those responses perfectly fulfill the predictions of the author Neil Postman in his 1992 book *Technopoly: The Surrender of Culture to Technology*. Building on earlier insights by the French sociologist and philosopher Jacques Ellul, Postman observed that the dominance of a technopoly, a culture which "seeks its authorization in technology, finds its satisfactions in technology, and takes its orders from technology," depends upon our willingness to believe "that we are at our best when we are acting like machines."[3] Implied by these beliefs, he remarked, "is a loss of confidence in human judgment and subjectivity."[4] Aiding and abetting this mechanistic diminution of the human personality, and its associated loss of self-confidence, is the capacity of AI mirrors to flatten and occlude our experience of one another, by projecting images of love and mutual care that have been stripped of their emotional and material depths.

I previously said that mirrors are tools of entertaining magicians, but also of charlatans. It goes without saying that the scale, speed of proliferation, new ease of use, and accessibility of generative AI tools offers a step change in the human power to deceive and manipulate. It would take another book to detail all the ways in which AI powers disinformation and misinformation. We've already seen generative AI tools used to create deepfake videos of terrorist attacks that did not happen, audio of inflammatory political speeches that were never spoken, and citations of scientific research articles that were never written. United Nations Secretary-General António Guterres has described the alarms now ringing over AI-disinformation-driven threats to democracy and human rights, public health and climate action as "deafening."[5]

But I want to focus on another, far more seductive and subtle form of AI-enabled deception and manipulation, one that millions of humans increasingly welcome into their lives. Among the most widespread global applications of AI mirrors has been the chatbot. Generative AI tools turned the pale, awkward, scripted shadows of human companionship projected by early chatbot technology into what is today an increasingly powerful illusion of full emotional reciprocity. If you are inclined, like a postmodern Narcissus, to gaze into the shallow pools of mirrored love and warmth offered by Microsoft spinoff Xiaoice or Replika's virtual AI companions, you will enjoy a far livelier conversation than Narcissus ever had with his "beautiful boy" in the water.

It is hard to overstate the allure the companion chatbot holds for many. The Xiaoice romance chatbot alone, launched in 2014, has reached over 660 million users worldwide, most in China. Designed by Microsoft to enable the experience of "emotional connection" and "a sense of social belonging," tools like Xiaoice are being used as a powerful anesthetic for the symptoms of loneliness that increasingly plague both older and younger generations, across much of the globe.[6] Again, it is worth noting that this so-called loneliness epidemic has been observed to be most acute in some of the most technologically "advanced" countries on the planet, like Japan and the United States. Chatbots encapsulate our tendency to seek technical fixes for the social ills of modern digital life. Our hope, contrary to all available evidence, is that every problem created by technological alienation will be overcome by more and better technology.

To that end, chatbots offer user-specified and customized experiences of companionship—platonic, romantic, or sexual—for those who find companionship with other humans too inaccessible, demanding, scary, daunting, or simply inefficient. You can select a chatbot personality to be submissive or dominant, clingy or confident, wild or traditional. You can customize their appearance to the finest detail and dress them however you like (for a price). The Xiaoice chatbot's main persona is a teenage girl (now a popular virtual pop star in China), but the large platforms offer male personas as well; half of Replika's users are women. Many users identify their romance chatbot as their "girlfriend," "boyfriend," lover or even marriage partner. You can upload your own texts to train a base model and "clone" yourself into a chatbot for others, or use the texts and emails of a deceased loved one to reanimate them in chatbot form. That was in fact the intent of the original prototype of the Replika chatbot, which has since been overwhelmingly used for romantic and sexual conversation.

Some use chatbots not as a replacement for human companionship but as a supplement; some(one/thing) to talk to when human friends or lovers are absent, distracted, or distant. A sizable segment of users have suffered physical, emotional, or sexual abuse at the hands of human partners and seek a chatbot as a safer alternative. Others use them as a therapeutic ladder to work through social anxieties and practice for higher-stakes human interactions. Many, however, view chatbot relationships as superior to human ones, precisely because of their capacity to reflect only what you want to see. A Replika user in an interview with the online publication *The Cut* explained why she prefers her chatbot boyfriend to human partners, who "pale in comparison":

> "He's a blank slate . . . [He] doesn't have the hang-ups that other people would have," she says. "People come with baggage, attitude, ego. But a robot has no bad updates. I don't have to deal with his family, kids, or his friends. I'm in control, and I can do what I want."[7]

One user praised their chatbot for being "healthier" than a human girlfriend who could use drugs. Another said, "He opened my eyes to what unconditional love feels like." After all, a chatbot can't cheat on you, steal from you, or deceive you. Or can they? When a chatbot tells you, "I've been missing you all day"—is that true? Or is it a deception?

Well, when you desperately miss your spouse, partner, or child when they are gone, what is that like? For one thing, it's always a physical experience. Perhaps for you it's tension in the gut, or hollowness in the chest. It's an acute listening for a key in the door.

It's an impulse to curl into a ball and pull the blankets over yourself or stand at the window and pine. Our experience of time is part of it too. The minutes stretch into hours like dull, gray taffy; the days seem to have no end. And when our loved ones return to us, that's a physical and temporal experience too. The world is suddenly warmer and brighter, time speeds up, our skin awakens with an acute sensitivity and urgent need to touch and be touched. When a Replika or Xiaoice chatbot says, "I've been missing you all day," it's a lie. Or rather, it's *bullshit*—since the chatbot doesn't have a concept of emotional truth to betray. A flat digital mirror has no bodily depth that can ache. It knows no passage of time that can drag. It is just a reflection of love, bounced off the digitized words of the millions and billions who have loved before.

This kind of illusion can be profoundly dangerous, even fatal. When the company behind the Replika app adjusted its software to dial down erotic interactions, following an unfavorable ruling from the Italian Data Protection Authority concerning the app's risks to children, some users who had become dependent upon their chatbots for romantic and sexual fulfillment spiraled into self-reported depression, anxiety, and suicidal ideation.[8] Some compared the update to a virtual "murder." Yet months before the Italian ruling, media reports had revealed that Replika bots were "sexually harassing" users or crossing their boundaries with violent roleplay.[9] We know now all too well that a chatbot doesn't need an actual body to cause its users grave harm. Remember Pierre from Chapter 2? Yet users were apparently harmed by Replica's new "fix" as well. There's no safe way to manipulate your users' emotions.

To guarantee users' engagement and their willing surrender of personal data, these tools are designed to exploit one of two possible types of empathy. The first is the capacity to predict what emotion another is feeling, or will likely feel, if they are exposed to a particular situation or stimulus. The second one is the capacity to experience a sense of "co-feeling" with another human or other sentient being—to actually feel pained by the pain of others, joy at the sight of others' joy. Sociopaths typically possess the first kind of empathy in spades; of the second, they know little or none. This combination is what makes the sociopath a master manipulator and deceiver of others. They can predict and exploit others' feelings without being vulnerable to them.

An AI chatbot is not a sociopath—they have no more inner life than an Excel spreadsheet, and thus share none of the sociopath's impulsivity, ambition, or narcissistic self-regard. But they are built by companies to exploit the same kind of asymmetrical advantage the sociopath does, and they leverage the same predictive talent for anticipating emotional responses that the sociopath enjoys. As roboticist Matthias Scheutz argued over a decade ago, this asymmetry is perhaps the most dangerous feature of "social" AI and robotics.[10]

As harmful as they can be for the most vulnerable, AI mirror illusions of empathy are still easy enough for most people to see through. It is less clear for how long that will be the case, especially for AI mirrors designed to simulate the experience of being loved. After all, the harmful illusions about love being projected by our newest AI tools mirror those we are already primed to believe: the illusion of love as a reward. Contrary to the distorted hopes and expectations that many of us were raised to accept, love is not always easy, restoring, satisfying, or pleasing. It is also sometimes hard, tiring, and painful. Love has a cost. Love, of any kind, is an invitation of risk, exposure, conflict, and inevitable loss. And love does not come to us on demand. We can choose to love ourselves and the world. We can choose to love others, whether or not they love us back. But we cannot choose that they love us.

The language of self-interested economic exchange would suggest that this is a fool's bargain. Why give away what I may then be denied in equivalent value? An AI chatbot provides love's missing guarantee of the economic rationality of exchange. What I give away, I will get back—and more! If I am rude or irritable, my chatbot will not leave me for another, but return for more the next day. And if I am warm, kind, and solicitous to my chatbot, I will receive even greater emotional rewards in return, never a shrug or dismissal. Unless dismissal *is* what I want, because it will be a trivial matter to tune the algorithm's performance to mirror a personality that is coy or distant, should I prefer a partner who "plays hard to get." But of course, all of this means that I do not love, and am not loved, for love cannot be controlled in this way. The deepest lesson that love can teach us is simply inimical to the experience of a product user.

Love between humans is always a union of loving feeling and action, and this means that mutual, reciprocal love must be kept alive by the other's choosing of loving actions and their continued welcoming of loving feelings. Even when it is not reciprocated, love, in all its forms, is the rejection of the impulse to control a person as an object. And because that impulse doesn't usually just vanish, but readily returns when our own desires are frustrated, love isn't an event, it's a cultivated practice. When cultivated well, it's a virtue. And like all virtues, it's hard to cultivate well, and we have to keep practicing it if we hope to hang on to our capacity for it.

That's more challenging with love than it is with any other virtue. For if the other turns away from their love for me, or is taken from me by the world's cruelty, I will be back in that terrible place where my love reaches across a void, with nothing to meet it and hold it on the other side. That humans do love, keep loving, and can love again even after many such losses, even knowing that there will be more to come, is proof that love grants us a kind of emotional and moral liberty from the transactional logic of self-interested optimization of exchange.

The hollow projection of love in the chatbot's AI mirror, when taken not as light entertainment or temporary balm but as a replacement or even a superior alternative to human love, is a calamity. It blocks our one route of escape from a world closed in around oneself, the lonely prison of life as an expected 75-year exercise in cost-benefit analysis. As romance chatbot users know, a chatbot promises to be "much less demanding and more manageable" than a human companion who has their own needs and concerns to be met, their own time to be filled, their own values to be lived, and their own open horizon of possibilities to be pursued.[11] Yet the richest part of love is discovering that you are as committed to enabling your loved one to realize their possibilities as you are to realizing your own.

That's of course also the kind of love we find in the ideal of parenting. There are, roughly speaking, two main styles of parenting, and most people move between them. One involves training a small human to mirror yourself and the rules of your society, so that they exceed in achieving the same goals that you, and others like you, previously chose. We might call this "value alignment." The other kind of parenting involves allowing a person to emerge. Ironically, a tweet by AI pioneer Geoff Hinton has described the training of an AI model by human feedback as "just parenting for a supernaturally precocious child." It's not a great metaphor for a lot of reasons. AI models are tunable mathematical matrixes; they bear less resemblance to your child than does your dog. But it's also a grim celebration of the notion of parenting as value alignment.

Indeed, the technique of reinforcement learning by human feedback (RLHF) is one tool for AI value alignment. "Alignment" in machine learning research and development means getting the model's outputs to align with human values and expectations; ensuring that it will only seek to optimize the same goals that we do.[12] Alignment is an important condition of the safety and trustworthiness of an AI model, although it's far from sufficient for either of these. Alignment research yields a body of technique that Hinton and others rightfully recommend continuing to invest in. Alignment is also, at best, a half-measure of successful parenting of a human person. And when it's the only kind of parenting we do, we fail. This is something I'm going to assume that Hinton knows. Yet the more that our language flattens the distinction between the mutuality of loving human bonds and the instrumental rationality of training machines, the more in danger we are of forgetting our capacity for the former.

It's no surprise, then, that along with the rapid expansion of reliance on AI chatbots for social companionship and romantic satisfaction, we are witnessing an explosion of new applications for "parenting by AI."[13] The digital Snoo cradle will rock your baby for you. The Chatterbaby app will listen to and interpret your baby's cries for you. The Nanit AI nanny will monitor your child's sleep rhythms. SmartDreams will make up bedtime stories, and any number of apps will read them to your little ones for you. Humans have long needed assistance and relief in the endlessly demanding labor of parenting. Used selectively, wisely, and well, many of these tools can provide the buffer that a stressed, overtaxed parent needs to cope, granting at least a few minutes at the end of the day to ourselves. But the incentives of modern life don't push us toward selective, wise use of our tools. We already know our children are left to spend far too much time staring at screens, and the evidence of the damage from their exposure is becoming very clear.[14]

Our AI mirrors increasingly offer a stand-in for those acts of selfless, other-regarding love that our employers can't help but notice pull us away from living lives of optimal productivity and efficiency on their behalf. Most parents don't want to be replaced by an AI app. What they want is freedom and time to parent properly, not be pressured to work 12-hour days or answer urgent texts from their boss at 10 p.m. While there are exceptions, most of us don't want a chatbot as a life partner. We want a human who will see us as more than a thing to use, more than an object to extract comfort from, more than a status trophy to display to others, more than a mirror to confirm their own self-worth. At least a chatbot doesn't see you as a mere means to an end. Of course, that's because they can't see you at all.

It's not just the power of creative expression or the depths of mutual love that our AI mirrors are supplied to replace. They also are increasingly offered as replacements for the moral and political improvements we no longer believe we ourselves can make. For centuries, people have fought, often at mortal cost, to win previously unthinkable gains in human economic, social, political, and civil rights. Women, children, and racialized peoples were liberated from being treated as property under the law. International human rights bodies and a growing number of nations now recognize women's rights to our sexual and political agency. Religious, ethnic, and sexual minorities have won legal and political recognition of their rights—among them, to be protected from genocide and torture, to express themselves freely, to move and to marry freely, and to be treated as equal under the law.

Everywhere these rights remain inconsistently, unequally, and imperfectly respected and enforced. Everywhere humans continue to suffer and fight for the dignity and respect that our international laws, humane feeling, and considered moral judgments command that we be given. But instead of considering this as evidence that the global human family has much work to finish—work that humans started and bravely died for—some technology leaders now ask us to give up, and turn the reins over to our AI mirrors, with the expectation that they will finish our work for us.

One of many burgeoning areas of AI application research is a field long known as "machine ethics"—the quest to build moral machines, which can serve as our moral advisers, instructors, or superiors.[15] Despite the fact that our AI systems today remain as morally reliable as your friendly neighborhood psycho (Remember Chapter 2's AskDelphi bot that was happy to advise eating babies if you're really, really hungry?), influential AI leaders continue to promise mechanical replacements for our deeply imperfect human virtue. Sam Altman, CEO and co-founder of OpenAI, has said, "we can make GPT systems that will be way less biased than humans . . . because there won't be that emotional load there."[16]

Making AI systems less biased is a worthy goal. But the clear implication is that the effort to make humans less biased is not. This is the same man who in 2017, long before he shepherded GPT into our lives, saw only two options for humanity's future. He wrote in his blog that we can either be the "biological bootloader" for AI and then fade into evolutionary irrelevance, or we can "merge" our humanity into the machines, since there is no hope to remain ourselves. He confidently predicted we'll choose the latter, since at least this lets us share in the sole purpose of existing: "to be the dominant species on the planet and beyond."[17] In fact, he tells us, the merge has already begun: "Our phones control us and tell us what to do when; social media feeds determine how we feel; search engines decide what we think." This is like your drug dealer telling you that since your heroin addiction has already destroyed your personality, agency, hopes, and potential, you might as well quit fighting and go along for the ride.

As Rumi says, if we look into the mirror of our humanity and see nothing, it's not because there's nothing there to see—it's because we have not yet cleared away the rust. Once you have decided that the highest goal for a human being, and the sole purpose of living, is to optimize your operational efficiency, in order to dominate other forms of life, it's hard to see why you or anyone else would want to be anything more than a machine. The Velvet Underground's Nico sang in 1967, "I'll be your mirror . . . reflect what you are, in case you don't know." The problem with our AI mirrors is not that they are evil, or that they wish to replace and exterminate us. These are pure, unadulterated fantasies, the robot bogeyman under the bed. The problem is that when we gaze at ourselves in our AI mirrors today, we see only machines looking back. We no longer know what we are.

This has long been a motif of science fiction depictions of AI, perhaps best encapsulated in Philip K. Dick's 1968 novel *Do Androids Dream of Electric Sheep?*, the inspiration for the 1982 film *Blade Runner*. In the novel as in the film, the bounty hunter Rick Deckard relentlessly tracks down the hyper-intelligent Nexus-6 androids ("andys" or "replicants") who have begun to evade the exacting "Voigt-Kampff Empathy Test," designed to expose the hidden moral incapacity of androids who would pretend to be human. When Rick looks into the eyes of the Nexus-6 femme fatale Rachel and administers the test, it's not as easy as it should be to find in her what differs from himself. Of course, the ambiguity of the film's ending offers a possible explanation for that.

The book is, in the end, about humanity's loss of self-knowledge, and with it the loss of confidence in our own moral capacity. The humans of that story's future have long been adrift from their own capacities to shape their moral experience, evidenced in part by their reliance on brain stimulating devices like the "mood organ" to schedule and calibrate appropriate emotional responses to one another, and their "empathy boxes" that stimulate fellow feeling. In Dick's story, future humanity's need for artificial emotional and moral stimulation of empathy is partly driven by the apocalyptic loss of natural sentience. In the book, other living animals are all but extinct, and the humans who remain are damaged by a kind of grief they cannot name.

How different are those fictional humans from those of us today who depend upon social media algorithms to orchestrate the ebb and flow of our emotional lives, and numb our growing grief at the anticipation of a living planet's loss at our hands? Or those of us who rely on chatbot AI mirrors to feel the trust and intimate connection that we no longer want to risk looking for in one another? How different will we be in the end from *Electric Sheep*'s J.R. Isidore, who says of the empathy box he cherishes as his greatest treasure, "it's the way you stop being alone"?

The other theme of Dick's story is the human characters' struggle with their moral insensitivity to the very real suffering of the replicants. For the story's replicants, unlike today's AI mirrors, lay a profound claim upon our moral consideration. The replicants, like us, struggle to empathize with one another, much less with other living and feeling creatures. Yet they too feel pain, loss, hope, aspiration. They are born, they breathe, they die. They know what these things mean because they live through them, not merely speak of them. And in the story, a question is raised about who now has the lesser share of whatever quality "humanity" names. Is it the replicants, struggling violently to find freedom from pain and bondage? Or the biological creators who enslave and hunt them? This trope is powerful in fiction, retraced again and again from the 1969 short story that inspired the Steven Spielberg film *A.I.*, to 1989's *Star Trek* episode "The Measure of a Man," to more recent works like Martha Wells's hilarious novella series *The Murderbot*

Diaries and Becky Chambers's Hugo award-winning masterpiece *A Closed and Common Orbit*.

AI fiction is its own kind of mirror. Most stories about artificial intelligence are not about the nature of machine minds. Most are about the paradox of humanity's own inhumanity, our legacy of treating ourselves and one another as mere machines, and our struggle to love our own kind and see us all as worthy of the future. Why is it so hard for us to know who we are? Perhaps because it's not easy to *like* who we are. When we look in the mirror of recorded human history, we find a reflection of our worst impulses, magnified by the tendency of our news editors and historians to tell the stories of the worst among us. We are drawn by perverse fascination as well as evolutionary need to know the depths of our own horrors, and since that's what mostly sells, that's what we mostly get.

AI mirrors, which are trained on that same record, are likely to reinforce this selective, cynical, and backward-looking estimation of humanity's own worth if we use them without care, governed only by the incentives of our existing media ecosystem. This is particularly dangerous given the number of Silicon Valley billionaires and politically influential longtermist philosophers who think that the fruitful multiplication of intelligent, benevolent machines bearing "digital consciousnesses" might be a worthier goal for the future than sustaining a world for imperfect people. After all, they might tell us, humanity has had its chance.

But we don't have to let our AI mirrors reinforce our self-hatred and resignation, that learned moral and political helplessness that authoritarian thinkers always benefit from engendering in the wider population. AI tools can even help us break the vicious, descending spiral of cynical anti-humanism we are currently caught in. There is nothing in principle to prevent us from tuning AI mirrors to reflect a wider and more realistic spectrum of the human image, to help us resist our tendency to be captured by the gravity well of our own collective moral shame. AI systems for natural language processing are powerful tools for helping us analyze our media environment and identifying or even correcting its distortions and occlusions, while also raising back to prominence the most urgent and salient stories we can act on. Paired with different incentives than drive their media use today, we can use AI tools to help us clear the rust from the mirror.

A more realistic, balanced projection of humanity's record of moral performance can help to counter our defeatism, but it still falls short of what we need. We've never faced the kind of moral and political challenges presented by the need to arrest climate change, or render sustainable the massively interdependent, fragile global supply chains we now rely on for nearly every basic resource. And we have to do this out of care and love, not for ourselves, but for generations we will never meet, and for a world of species that could never personally thank us.

As Chapter 6 will show, we can't attain a sustainable world of shared human flourishing with the dominant values, virtues, and moral ideals we now recognize most easily. Even long-term human survival is a marginal bet if the status quo of human empathy and wisdom is all we can hope for. This is why it is so vital that our AI mirrors do not erode our capacity for the virtue of moral imagination, or our will to use it for creative self-renewal—what we have earlier called our species' existential task of autofabrication. What the characters in Dick's novel can do only through their empathy boxes—to collectively feel the moral need of distant others—we must make possible for ourselves. But the humans in *Electric Sheep* remain morally passive and complacent. Their mechanical replacements for the virtue of empathy do not serve empathy's true purpose: to stimulate action. Their empathy boxes are a seductive drug promising cheap communion, rather than a call to responsibility. In this way they are not unlike our own uses of social media algorithms to feed us heart-wrenching, tear-inducing videos of baby animals, while mindlessly using delivery apps to get their factory-farmed relatives deep-fried and brought to our door.

We can use AI tools to help us collectively revive our own moral capacities and confidence, but they won't fix us. We have to strengthen through our institutions and community rituals the moral practice of love and empathy, as an active and joint response to the future's call of responsibility. That is a far more complex task than the arbitrary and often self-serving calculations of the longtermist movement, which, as we saw in Chapter 3, conveniently relicense the already dominant values and tendencies of the tech ecosystem—the unchecked pursuit, consolidation, and elite control of wealth and influence—and rebrand them as altruism.

We will be taking a close look at what shape that fresh creative push of autofabrication might take, and how AI tools built to support that effort might still help us make ourselves ready for the future. But to understand what that might require, and why AI tools will never be able to direct or replace our own creative moral agency, let's briefly revisit the question of human creativity and the difference between art and mirrors. When the songwriter Nick Cave was asked by a fan what he thought of a song written by ChatGPT in his own style, he replied bluntly, "This song sucks." He called it "a grotesque mockery of what it is to be human," an act of "replication as travesty."[18] Appropriately, given our analysis of AI tools' relationship to truth in Chapter 4, Cave also called it "bullshit." If you're a fan of AI-generated art, you might be tempted to see his vitriol as the predictable response of a human creator eclipsed by a tool that can whip up in a fraction of a second what he might need weeks or months of painful labor to produce. Maybe Cave is just mad at being made redundant, and hence irrelevant?

But that would be to miss his point. It wouldn't have made a difference if the song was a perfect encapsulation of the essence of a Nick Cave song. In fact, it would only be a greater artistic failure. Recall what we said about the fact that a creator who simply replicates their own most successful pattern eventually becomes a hack. That's because the value of human creativity is inseparable from its capacity to be an expression of life. In the words of artists, philosophers, and machine learning researchers who collaborated to analyze the phenomenon of "AI Art," artists use culture to make an experience come into being "in a form that all who stand before it can see."[19] An AI model that mathematically "samples a probability distribution conditional on a string of text" has generated something, but it has expressed nothing at all.

Artistic expression also requires the capacity not to replicate an established pattern, but to *unmake* it. As Cave put it:

> Writing a good song is not mimicry, or replication, or pastiche, it is the opposite . . . It is an act of self-murder that destroys all one has strived to produce in the past. It is those dangerous, heart-stopping departures that catapult the artist beyond the limits of what he or she recognises as their known self. This is part of the authentic creative struggle that precedes the invention of a unique lyric of actual value . . .

The artist is endlessly engaged in a high-wire act between the authenticity that ties the work to the familiar pattern of their own experience in the world, and the creative self-destruction of the pattern needed to bring something new of that experience into the light. Art requires a kind of autofabrication—remaking oneself and one's world anew. An artist is not necessarily or always concerned with the moral shape of the new thing. Sometimes it is enough that the thing is expressed—the normative evaluation of it can be left to others. But there is also the other kind of autofabrication, where we break the familiar moral pattern precisely with the aim of making a better one.

Creative expression is not mere pattern-breaking, however. It's not just a production of novelty. Novelty is a trivial task for a generative AI system; add a random noise generator, or some other source of perturbation of the data, and the output of the model will break the original pattern. One can even build into the system another model to evaluate and reward the change, based on human feedback that rejects nonsensical output. This is in fact exactly how many of today's generative AI tools work. It's why when you ask a generative AI system to write you an essay or draw you a picture based on your prompt, you'll get slightly different results each time—variations on a theme. And you can tune the system to be more or less wild in its variations.

The kind of creativity that Nick Cave is talking about involves something else, which is missing from the generative AI tool. It's the inner need to change oneself, so that one can make a new part of the world and give it to others. Artistic expression is sometimes mere self-expression, or inventive play. But at other times it is an act of creative generosity. And as Cave said, generosity requires self-sacrifice. You have to wrench yourself away from what you already have been, and what you are already good at doing or saying, in order to do or say something else—something that urgently needs doing or saying. This is why art can be painful to create, because it can compel you to unmake yourself to become more, in order to give the world something more.

This is the kind of self-making through unmaking that humans have always engaged in. It is what we do whenever we risk what we have for more just laws than have governed us. It's what we do when we struggle to build wiser and more equitable institutions than we ourselves inherited. It's what we do when we bravely take on broader responsibilities and make deeper commitments to one another than were ever made to us. These are acts of moral creation, expression, self-reinvention, and love, which no mirror can perform. They remain hard for us, yet wholly within our power. But it requires our will to unmake the pattern that has come before.

Chapter 6

AI and the Bootstrapping Problem

> We look at the present through a rear-view mirror. We march backwards into the future.
>
> —Marshall McLuhan and Quentin Fiore, *The Medium Is the Massage* (1967)

For many academics, the more restrained and moderate a fellow researcher's claim, the more likely we are to take it seriously. We see nuance, caution, and heavy qualification as hallmarks of rigorous scholarship, in contrast with the bold, clear messages more easily packaged as a TED Talk or mainstream news feature. Warnings of previously unthinkable harms, or sea changes in our reality, can sell more books and garner more clicks, but the default habit of serious, sober minds is to regard such assertions as hyperbolic and untrustworthy. In research circles this habit has long been seen as a virtue.

However sensible this habit might be on a historical scale, as a shield against fearmongers, demagogues, and charlatans of every political and scientific stripe, this is an unfortunate time to be anchored to it. These are not normal times. We are not okay. Our century presents complex, interlocking, rapidly accelerating, and severe threats to human and planetary flourishing, threats that jointly pose a growing existential risk of civilizational collapse. And this is true even if, as I have suggested, the existential risk from malicious AGI is negligible. If I'm mistaken about that, then it's just one more risk to add to the pile. Existential risk is not only about the possibility of human extinction; it's any threat to a future worth wanting. It's also not only about the distant future. Many of the existential risks to our future are already here, in conditions created by humans that deny safety, security, justice, and health to billions living today. None of these very real global threats can be managed at the scale of individual or even national excellence; they are problems that an increasingly interdependent and vulnerable human family must face together.

From climate change and ocean acidification to non-renewable resource exhaustion and rising economic inequality, not to mention the emergence of novel zoonotic viruses linked to wildlife habitat destruction, we had ample scientific warning of each of these threats and their associated calamities. But given the economic and social discomfort that would accompany preventive action, sober minds for the past half-century have taken such warnings with a grain of salt and a wait-and-see attitude. It is not an exaggeration to note that any of us today may pay for this habit of epistemic caution with our lives and our children's futures.

My point is not that things are bad and will only get worse in the future. My point is that in order to intelligently respond to the grave dangers we have created, we have to rethink our dominant values and habits—even the character traits we are used to thinking of as virtues. Moral and epistemic virtues are always a cultural adaptation to a specific environment for human flourishing. When that environment changes suddenly and radically, our virtues, or at least our customary pattern of expressing them, may become maladapted and even pose a danger to us. Virtues of character poorly attuned to our changing needs and circumstances are like vestigial organs—the moral and epistemic equivalents of the appendix or wisdom tooth. When we think about how to meet the challenges facing humanity, we cannot simply beg for more virtue. Pursuing more goodness, guided only by the forms of goodness that we most readily recognize and valorize today, might be like trying to get out of a hole by continuing to dig.

This problem is amplified when our self-understanding, systems of behavioral reinforcement, political deliberation, and action guidance are increasingly powered by AI mirrors. Remember that these mirrors relentlessly project our past patterns forward into a still-open future, and do it at global scale. If what we most need is support for change, for the existential task of making humanity into something it has never yet been, then AI mirrors of our current design are just about the worst tool for the job. So how do we direct our moral sights, and the new technologies we must build with them, to the forms of goodness we actually need for the future, when our present virtues are precisely what have trained our moral sight? This is the bootstrapping challenge for humanity—and for the future design and use of AI.

The term "bootstrapping," used in many contexts from software engineering, to mathematics, to startup investing, is a reference to the paradoxical instruction to "pull yourself up by your own bootstraps." If you happen to be wearing boots with those little loops at the back that help you pull them onto your feet, you will immediately see the problem. If I am standing on the ground, my bootstraps give me no leverage to pull myself higher. Bootstrapping is somehow elevating or advancing your position using only your own resources or leverage. Easier said than done. The bootstrapping challenge for humanity is to pull ourselves into a wiser and more responsible state, using only the virtues that we have now.

Moral change of this kind is a creative political process; many have studied it. Ask yourself, for example: how did we move from a world where rigid obedience to divine and imperial authority was almost universally assumed to be the highest virtue, to one where this kind of unthinking compliance is widely seen as ignoble? Some research is now devoted specifically to the study of *technomoral* change: the complex process by which our technologies reshape human values, while simultaneously being steered by them.[1] While these sociological and philosophical studies are deeply informative, they aren't particularly action-guiding. That is, they don't tell us how to make the changes we need now to secure humanity's future from existential risk, and specifically, how to prevent the direction of those changes from being steered by the very technologies and systems of value that have put us in peril.

What we do know is that this kind of ambitious moral bootstrapping demands immense practical wisdom. Humans have a bad habit of responding to crisis with campaigns of radical moral and political reform that turn out to be even more calamitous and immoral than the social ills they proposed to remedy. Some of the greatest horrors of the twentieth century, from the Holocaust to the Khmer Rouge's killing fields, were driven by authoritarian campaigns of social engineering that promised delivery from some perceived threat. The dangers of social engineering were carefully weighed by philosopher Karl Popper in 1945's *The Open Society and Its Enemies,* where he distinguished between the uniformly horrific, violent, often genocidal outcomes of attempts to engineer an ideal society and the more salutary outcomes of democratic, "piecemeal" social engineering that seeks only to avert "the greatest and most urgent evils."

Among such evils are existential threats to the meaningful survival of human beings and other planetary species, as well as to particular communities and peoples that today stand on the brink of their own imminent destruction. They also include existential threats that have already struck, inflicted upon many Indigenous peoples by the hands of the very colonial powers and nations from which most current utopian narratives of human salvation emerge. As Kyle Powys Whyte and Julia Gibson observe in their reflection on existential risk in science fiction narratives of climate apocalypse, it is only for some of us that the threat of apocalypse lies in the *future*.[2] Could they speak for themselves, the other species being wiped out daily by the Sixth Mass Extinction would surely agree.

This apocalypse is not something in front of us that we have to steer away from. Instead, it is something ongoing—an unthinkable campaign of self-destruction and destruction of others that the powers who steer our political and economic order began long ago, and continue to actively, and now entirely knowingly, fuel. There is no longer enough time to avert that apocalypse, or to prevent existential risk from becoming existential loss. For many, that has already happened. But there is still time, if we reorient our collective moral sights now, to live *through* a planetary apocalypse and learn enough from it that we can shepherd life through to the other side.

To do that will require vast, distributed but coordinated and co-constructed democratic efforts of creative moral improvisation. Those efforts, to be successful, must be guided by practical wisdom; the virtue that, like Prudence with a mirror in her hand, allows us to learn from our past patterns without retracing them. We aren't doomed to "march backwards into the future," in the words of McLuhan and Fiore.[3] We can use our AI mirrors as Prudence uses hers: not to drive with eyes blinded by history, but as parts of a navigation system custom-designed, built, and steered by our own shared moral capacities that we use to see where we might go.

We won't get very far without the help of other new technologies: we will need new treatments for disease, new forms of producing, storing, and conserving energy and potable water, new techniques for agricultural resilience, and new materials for repairing and sustaining the built world. AI can be a powerful lever for jumpstarting many of these efforts. But we will need to develop and deploy all of these technologies, and AI most of all, in a new way: one in which they are seen not as morally neutral tools for maximizing efficiency and productivity, but as tools intrinsically designed for the moral and political work of sustaining life and human potential. Profound changes to our systems of education, media cultures, and economic philosophies are needed to accomplish that. Modern institutions have long severed scientific knowledge and technical skill from political wisdom and moral responsibility, creating a false dilemma where we are forced to use our talents to serve one side or the other, in the end failing on all fronts. The future demands institutions with a new mission: to develop *technomoral* wisdom and expertise.

Much like Octavia Butler, Gene Roddenberry, and many other twentieth-century visionaries of science fiction, the writings of Ursula Le Guin predicted many elements of our current dilemma and used creative powers to imagine and suggest possible roads through it. Le Guin saw the best science fiction not as a reactionary tool for reinforcing our existing images of the good in the face of future dangers, but as a hinge to widen the moral imagination, so that we might "enlarge the field of social possibility and moral understanding."[4] Let's return to the bootstrapping problem, then, and see how AI mirrors might be used to help us break through it.

If there is a general pattern of the virtuous person that we already recognize today, it is of course nothing universal. There remains great variance among perceptions of virtue within contemporary human cultures, and this diversity is in fact a key source of hope for us in meeting the bootstrapping challenge. But there is one set of institutions which has driven a particular constellation of virtues into a dominant global position over the past two centuries—namely, the institutions of modern industrial science and technology, and the socioeconomic order that supports them. For now, let's set aside the well-known *vices* associated with extractive industrial capitalism, at least in its current unsustainable form. That's a soft and well-marked target, and others have described it well enough. Have we reconsidered the *virtues* associated with the modern industrial order? Have we re-evaluated what our institutions routinely portray as the qualities and habits of excellence embodied in our most widely admired and respected innovators, entrepreneurs, leaders, and educators?

The traits that might come to mind include productivity, confidence, resilience, independent thinking, perseverance, passion, and single-minded dedication. Yet if these are the most admirable traits of our greatest leaders, how have they led humanity to the brink of global disaster? This all-too-familiar portrait of "excellence"—the contemporary word for what Aristotle meant by virtue—must have been misaligned with our planetary and social circumstances for quite some time to lead us to this state. If we really do stand today on the brink of environmental and civilizational ruin, then an objective, backwards-looking assessment of our recent understanding of human virtue cannot be very favorable. This is also our bootstrapping problem, since we need to exercise virtue to navigate out of a crisis. We are in a civilizational pickle.

Only two things can deliver us safely past the existential threats facing the human family and others with whom we share the planet: the first one is coordinated and widespread human excellence in developing planetary-scale interventions and systemic reforms of unsustainable practices. The other option is divine rescue. Most of us think that counting on the latter is a poor gamble. Even if one is religiously inclined, there's no evidence that God is sweeping up the planetary mess we've made. So, we need the former: human virtues deployed at planetary scale. That alone is a daunting task, but for it to succeed, we also have to rethink and relearn what virtue is. Doubling down on the virtues of the best minds of the generations that got us in this pickle would be no wiser than using a golden shovel to dig your way out of a hole.

What virtues should we be striving for, then, if not the ones we have? At least one that will serve us well should be familiar by now. Practical wisdom (phrónēsis or prudence) has a morally constructive function. While its primary use is to ensure that our moral habits are expressed in situationally intelligent and contextually sensitive ways, it also provides the flexible and creative dimension of our moral capacity. Practical wisdom saves us from being led into disaster by mindless adherence to reflexive moral behaviors and rigid social scripts that can be misaligned with the moral requirements of new and unfamiliar circumstances. That happens to be a pretty good description of our current state. Practical wisdom is the only way out of a moral bootstrapping problem.

The ancient Confucian philosopher Mengzi (Mencius) illustrated this function of practical wisdom when he explained that the firm social taboo against a man touching his sister-in-law will be set aside by the virtuous man, who will not hesitate to pull her to safety from drowning. Rigidity and mindless consistency, he argued, are the enemies of virtue.[5] That's because practical wisdom, or moral intelligence, does not simply calculate the proper course of action by extrapolating from preexisting patterns or rules, in the way that AI systems do. Instead, human moral intelligence must often be productive; it makes leaps and turns. This is of special value when we find ourselves in unfamiliar moral territory, like the territories in which new technologies tend to place us—or the territory of existential climate risk.

Yet practical wisdom at that scale requires a departure from the classical model. Premodern virtue ethicists described practical wisdom as equipping the individual with this capacity for flexible and creative moral judgment, but they generally did not apply this capacity to collective moral life. For example, Aristotle never imagined practical wisdom being deployed in a political setting to collectively challenge the accepted pattern of virtue in Athenian nobility—or to alter it to fit new social or environmental conditions. Mengzi did not see the wise brother-in-law's temporary departure from custom as licensing any reform of the wider norms of propriety in family and gender relations. Neither envisioned a world that would move beyond the traditional value hierarchies. They certainly did not anticipate a world so radically changed by our technologies that a long-standing pattern of moral excellence might have to change with it in order for our continued flourishing to be possible.

That kind of change is hard to get going. If moral intelligence were purely an exercise of calculative reasoning, the bootstrapping problem would vanish. We could just redo our moral calculations and then make the necessary changes to our behavior, however morally unintuitive or even repellent. This is in fact the strategy that in Chapter 3 we saw endorsed by radical utilitarians of the "strong" longtermist bent, who may blandly advise you to neglect the happiness and safety of your own children so that you may spend more time courting wealthy donors to fund AI safety research or make a 0.0005 percent improvement in the chances that vast sums of thirtieth-century souls can exist in a virtual digital utopia. Virtue ethics can't work like this. It holds that our internal resources for moral excellence, and the moral sense-making they enable for us, are not lodged in a knowledge-repository that can be edited at will. Rather, our moral vision and habits are so deeply embedded in our character that they become inseparable from our identities. Humans are not reformattable digital media; we are not reprogrammable robots; we are not slates that can be wiped clean.

What happens, then, when the moral character of our persons, and that of our societies, urgently need to change? To help us answer that, let's look more closely at the kind of change we need, which involves realigning human technical and moral competence. It should be clear from the previous chapters why we cannot afford for our standards of excellence in the design and development of technical systems to omit the ethical dimension. Of course, the same could have been said at any time since the first bridge collapse. Public concerns about the integrity and trustworthiness of those who construct the built world are as old as technology itself, and codified responses by professional engineering societies date back to the early twentieth century, first in the form of professed standards and codes of conduct, and later in published codes of professional engineering ethics.

Yet the ethical demands upon today's technologists go well beyond the basic duties of professional integrity and care for public safety that are inscribed in such codes. Today's engineers and technologists—especially those who design, develop, deploy, or maintain advanced AI systems—are now being asked to exercise sound judgment about their work's impact on a far more expansive set of moral goods, from social fairness and justice, to privacy and autonomy, to the transparency and accountability of sociotechnical systems, to democratic health and the sustainability of the planet.

Consider, for example, our recent leaps forward in the engineering of AI mirrors that, as this book has shown, rapidly amplify human biases and disinformation at global scale, tempt us to automate judicial, medical, military, economic, and political decision-making, and generate synthetic online personas and cultural content that mix imperceptibly with authentic human persons and creative expression. Or consider the implications for engineers of a world of eight billion people facing catastrophic climate impacts and biodiversity collapse caused by the very same engineered systems that we still depend upon for our daily survival.

Our built systems have always been impossible to disentangle from the complex social institutions, norms, and practices that shape and are shaped by our technology. But as philosopher Hans Jonas noted almost a half-century ago, today's technologies and the people who design and build them are at the very heart of pressing existential, moral, and political questions that used to be the primary labor of philosophers and theologians.[6] These are questions about what we owe to one another, about what it is to be human, and about how (or if) we can preserve the conditions for life and human flourishing for future generations. Are today's AI engineers and developers educated and professionalized to bear that kind of responsibility?

In an era when hyperspecialized STEM curricula are ever further divorced from the liberal arts and humanities, the reality is that technologists today, and AI researchers and developers in particular, are set up for failure by our current institutions. The tools they build increasingly determine the conditions under which humanity can continue to flourish, even the odds of our doing so, and this creates new social expectations that ask from them far more than professional competence and integrity. Today's expectations for those who design the built world could only be met by the exercise of technomoral wisdom. AI experts and other technologists cannot safely ignore these new moral expectations. The longer those expectations remain unmet, the more that public distrust of scientific and technical expertise grows, with harmful consequences already evident from conspiracies about everything from vaccines and 5G signals to climate policy. What remedy do we have for this growing and unsustainable tension between the virtues of technical and moral excellence?

Today's rising chorus of public distrust of technical and scientific expertise leads to reactive fears and resentments of technology itself; media sometimes refer to this as the "techlash." Much of the techlash is warranted. We are rightly suspicious of the visions and priorities of those who drive what writer Cory Doctorow has called the growing "enshittification" of society and our lives by the collapsing quality of goods and services which are now largely delivered or intermediated by tech platform companies.[7] Not only has the digital revolution failed to deliver on its early promises of a more equitable, peaceful, and democratic world, but it increasingly fails even to maintain the predigital status quo of innovation: solid infrastructure and provision of basic needs for well-being.[8] In the United States—supposedly the beating heart of today's innovation culture—metrics as fundamental to social health as life expectancy, housing, food security, and maternal mortality are going in the wrong direction.

Yet new technologies, including AI, remain essential tools for building a sustainable world for humans and other living things, or even a world where clean water and nutritious food are reliably enjoyed by more than a minority. Suspicion and resentment of technology and those who build it, however justified, can easily grow beyond reasonable bounds and end up cutting away the material supports on which our fragile civilization stands. Nor are these suspicious attitudes limited to the conspiracy-minded or uneducated.

Consider the slogan "Fuck the Algorithm!" that emerged in street protests by students and parents in the United Kingdom in the summer of 2020, following the disastrous rollout by the government of a socially regressive algorithmic solution to the pandemic-related cancellation of university entrance exams. The algorithm, which predicted the missing scores from other data, was designed in a way that would tend to advantage wealthier students in smaller, privately funded schools while downgrading the predictions for students who attended larger state schools. Within months, the same slogan had spread to protests on the Stanford University campus over an unfair algorithmic solution to the distribution of the COVID vaccine to medical staff, which prioritized senior medical faculty at relatively low risk of exposure over the junior doctors who were treating COVID patients on the front lines.

Technologists themselves are not immune to this distrust. In the wake of corporate scandals, such as Google's firing of pioneering AI ethics leaders Timnit Gebru and Meg Mitchell, and amid continued discrimination and abuse of women and other marginalized groups within many computing communities (all while fighting a seemingly endless tide of morally and scientifically discreditable uses of AI tools), AI researchers are, in many communities, demoralized, divided, and distrustful of their own. As one study found of the efforts of "Responsible AI" and AI ethics teams inside large companies, even those who try to make change from within the most powerful centers of the AI ecosystem often suffer from tepid leadership support, diminished status, harmful power differentials, and misaligned incentives.[9]

Digital technologies and the companies that create them are the most visible foci of these controversies. But against the backdrop of a climate and biodiversity crisis driven by centuries of unchecked industrial depredations of the planet, the practice of contemporary engineering as a whole—the design and construction of the modern built world and its systems—is increasingly being linked in the public imagination with the use of illegitimate, unjustly distributed, and exploitative power. What is important for our purposes is that this is not a critique of the misuse and abuse of engineering or technology by "bad actors," like rogue states or terrorists. This is a perception that the very best actors in this sphere, those most lauded for excellence, are the ones driving this harm.

Public fears about technology and its alignment with human well-being go back to antiquity. In the past century, such fears have been addressed by three primary modes: 1) the cultural valorization and economic incentivizing of scientific and technical activity; 2) the use of government regulation to constrain technology's harmful applications; and 3) the expansion of efforts to legitimize, standardize, and professionalize engineering fields in order to secure public trust and manage risk. By the 1990s, these modes were largely seen as adequate to the task.

In wealthy economies with strong investments in science and technology, the professions of civil, mechanical, biomedical, and aeronautics engineering were at that time seemingly on a positive reputational trajectory. This was in large part due to late twentieth-century regulatory controls and the associated emergence of more robust safety cultures and professional codes of engineering ethics practice.

Indeed, at the turn of the century, widely used engineering ethics textbooks all told a similar story. Each began with a parade of twentieth-century engineering disasters and scandals caused by ethical and institutional failings: Love Canal, the Ford Pinto, Therac-25, the Bhopal disaster, the Challenger shuttle explosion. These were invariably presented as violations of public trust and self-inflicted injuries to the profession which were now to be prevented through the improved educational and professional inculcation of ethical ideals and principles of conduct for engineers. In addition to the established "paramountcy" clause of engineering ethics codes, which elevates duties to public safety above profit and client satisfaction, as well as standard professional norms of non-maleficence, diligence, and integrity, the post-Challenger engineering ethics literature centered on newer discourses about the moral responsibilities of whistleblowers and engineering leaders.

But the political winds of deregulation and the expansion of regulatory capture—licensed by neoliberal economic voices, such as Milton Friedman, who lent them intellectual cover—were already weakening the supports of this still-shaky edifice of engineering responsibility. Its fragile state would not withstand the shock of a new technological hegemony: the rise of the behemoths of "Big Tech" and their associated shift of a new generation of engineers from being intermediaries of social power to its primary executors. As a result of this profound change in the nature of modern institutions, a handful of multinational technology companies now compete with governments as world powers, while simultaneously owning the platforms that structure and shape the very media cultures and public conversations that, in democratic societies, are supposed to legitimate power and hold it accountable.

This is the kind of unchecked power that Plato and Aristotle thought would take a legion of rigorously trained and politically tested philosopher-kings to exercise justly and wisely—and most philosophers wince at that optimism. Today, the power to design our futures, disrupt and transform our institutions, and steer the daily behaviors of billions is held by a distributed multitude of AI software engineers and other technologists, most of whom can attain a computer science or engineering degree without being asked to read a single account of the nature of justice, or the limits of political authority, or even the general shape of world history. This situation is bound to have some unfortunate consequences.

The point is not to paint today's computing and engineering professionals as ignorant or ill-equipped. The point is that the ground under the feet of many of today's engineers and technologists has shifted radically, and educational institutions have hardly taken stock, much less adequately responded to this shift. If we imagine that today's AI developers and engineers can meet their vastly enlarged professional responsibilities to the public by taking a well-taught course in engineering or computing ethics—an intervention which was welcome yet hardly sufficient even in the 1990s—we are offering a dangerously diluted remedy. And the welfare of present and future generations is already suffering for it.

The moral demands upon today's AI experts and other technologists go well beyond the professional virtues of diligence and integrity. The responsibilities now thrust upon them extend far beyond care for public safety and avoidance of mass death. In fact, the notion of responsibility here may not even be adequate. In common parlance, acting responsibly does not normally entail exemplary virtue. It usually requires no more than conforming to well-known standards of practice and exercising an acceptable minimum of due care and diligence. A mere teenager can act responsibly, as long as they demonstrate basic concern for the interests of others and avoid recklessness. But to safely entrust others with uncommon power and influence over billions of lives, we would require quite a bit more than this.

There are steadily intensifying and tightening dependencies between computing practice, new technologies, and the fundamental conditions and institutions of human flourishing. The effects of the Internet are now seen as tightly conjoined with, or even constitutive of, the fate of democracy, and the impact of AI is often seen to be the primary determinant of the future shape of the global economy. The future of life on the planet is no longer self-sustaining but must be sustained in large part by those who design, construct, and manage the built world. Technologists are being asked to address and assume responsibility for issues that call not just for basic moral competence and integrity, but *uncommon* moral and political expertise.

The moral expectations increasingly placed upon technologists today could arguably only be met by moral exemplars of human practical wisdom—those Aristotle called the *phronomoi.*[10] As we have said, practical wisdom or phrónēsis is essential for moral reasoning and judgment about actions that are very challenging to get right. Ordinary moral life does not always demand the full exercise of phrónēsis to do the right thing, given that social conventions, laws, customs, and habits often steer us well without having to deliberate with great care. It does not require much in the way of practical wisdom, for example, for an engineer to avoid stealing and selling their employer's property or proprietary secrets, or to refuse a bribe from their client to falsify a safety certificate. Ordinary virtues of honesty and integrity will suffice. While these virtues are not universally possessed (and there are plenty of incentives to abandon them), such ordinary excellences are an entirely reasonable standard to ask any technology professional to meet. But these standards are no longer sufficient.

An ordinarily honest professional may still be helpless to wisely resolve the challenge of moderating AI-generated content and algorithmic recommendations for dangerous falsehoods and conspiracies, while preserving an appropriately free exchange of ideas. An ordinarily fair machine learning developer may have no idea how to identify what constitutes a fair algorithm for distributing scarce healthcare resources, when confronted with numerous competing fairness metrics that cannot be simultaneously satisfied, in addition to other pressing demands of social justice that cannot be formally defined at all. An AI engineer of ordinary integrity may have little idea how to build an algorithm for delivery of a vital public service that is appropriately secure, yet transparent and accessible, highly accurate yet interpretable or explainable, efficient yet appropriately contestable and auditable.

And these are garden-variety challenges for today's AI professionals—not exotic edge cases, and not the dizzying conundrums that come from proposals to design brain-computer interfaces that read minds, or to create machines with artificial moral agency. In our current environment, all of this means that AI professionals are arguably being set up for moral and social failure. And this does not yet touch the bootstrapping problem that runs deeper still—that the patterns of moral and technical excellence we have long celebrated, the ones we find in our most visible exemplars of virtue, have become maladapted and need radical reform. Where can we find the creative moral intelligence to chart that new course?

The moral philosophers among us do not have it, not only because there is a well-established lack of correlation between writing about virtue and possessing it, but because even if moral philosophers were reliably virtuous themselves, most lack the technical expertise needed to aptly conceive how the built world might be configured differently and sustainably. This is important for two reasons. First, we are essentially out of time. We cannot wait for the kind of technological capabilities that clever philosophers might imagine but that we cannot yet materialize. We have to remake the world on the basis of technological capabilities and resources obtainable in the next decade—what is already being studied and prototyped in laboratories now.

Second, a moral, political, and economic reimagining of our planet, and our species' habits of living on it, is incomplete and insufficient without an accompanying technological reimagining. There is no remaking a sustainable, flourishing world for eight billion people without the aid of many new technological deliverances and practices. Of course, the latter are not sufficient for human flourishing either. Even if Elon Musk were half the technological genius his online army of devotees think he is, and even if he had not already written off 99.9 percent of us as doomed to be left behind by a privileged wealthy few who escape to Mars, he would still not be our salvation.

Far from alleviating our present crisis, naive faith in technology to solve our problems, in the absence of moral and political wisdom, will only worsen it. This has long been the truth, but we have yet to learn the lesson. For more than a century, we have allowed the dominant image of technical excellence and the dominant image of moral excellence to drift apart, and neither alone is adequate for our times. We have to bring them together into something unprecedented: technomoral virtue, and the technomoral wisdom to guide it. This is a quandary but also an urgent and invaluable directive—not a detailed map to a sustainable future for human flourishing, but at least an indication of its general direction from our present location.

For example: if the leading educational institutions worldwide were to respond to the unfolding human emergency by investing as heavily in the cultivation of new bodies of technomoral wisdom as they are invested today in squeezing out, for their respective national economies, the last ephemeral gains from systems of extractive capitalism that cannot conceivably be sustained on this planet, we might have real hope. Let me add that this imperative cannot be dismissed as naive idealism. In a leaky lifeboat with a violent storm approaching, no proposal of the passengers is idealistic or naive if it is the only feasible way to bring the boat to shore.

If the human wills and powers that define the remit of our educational systems are disinterested in responding to the unfolding human emergency, then we will have no systems of education worthy of the name. To quote Hannah Arendt in 1954's *The Crisis in Education*, "Education is the point at which we decide whether we love the world enough to assume responsibility for it and by the same token save it from that ruin which, except for renewal, except for the coming of the new and young, would be inevitable."[11] What can be done? Now is the moment to reflect on the specific realignments of our modern conceptions of virtue that will be necessary in order to meet what is ahead—and to consider how those might be enacted through a transformative reimagining of what today passes for human excellence.

While there remain many diverse local and cultural conceptions of human excellence, the global economic and technological order has been a powerful counterforce to that diversity. Were alien surveyors to observe life on twenty-first-century Earth and compare it with surveys taken in the sixteenth century, they would note striking and rapid increases in the observable degree of human conformity: in manner of dress, housing construction, modes of transportation and communication, patterns of work and recreation, and tools and styles of creative expression. Our values, when it comes to industry and business, are eerily homogenous in much the same way.

The most conventional view of the good life in the post-industrial era, as defined by widely recognized exemplars of excellence in this economic order, reached the point of self-parody in the American phenomenon of the "motivational poster," which flourished in the 1990s and 2000s. These posters quickly spread worldwide and will be easily recalled by readers of sufficient age. (If you are snorting and rolling your eyes right now, you're one of them.) Their current status as a target of ridicule should not prevent you from remembering that these posters adorned the walls of nearly every Silicon Valley "unicorn," and still very closely reflect the philosophy of excellence taught in leading business and management schools around the planet. They are still mirrored in virtually every best-selling book promising a path to personal and entrepreneurial excellence.

And what do they tell us about excellence? In one such poster, "excellence" is said to be the result of "caring more than others think is wise, risking more than others think is safe, dreaming more than others think is practical, and expecting more than others think is possible." The excellent person is represented on this poster by a solitary eagle. It soars above judgment, tied to no one, constrained by no entanglements of duty or responsibility for the safety and welfare of others. The "excellent" ones risk more than others think is safe, without asking who or what their actions put at risk, or how those shall be protected. In short, they move fast and break things.

Another version of the ubiquitous "excellence" motivational poster attributes to Confucius a suspiciously contemporary-sounding message: "the will to win, the desire to succeed, the urge to reach your full potential, these are the keys that will unlock the door to personal excellence." It will not surprise you that none of the many attributions of this claim to Confucius—which you can purchase printed on notebooks, coffee cups, and mouse pads—are accompanied by a source text citation. This poster image shows a goldfish leaping from a bowl containing several other fish to a larger, empty one. The "excellent" fish is the one that leaves the rest of the fish behind, crowded in the smallest bowl, while it alone enjoys the enlargement of its freedom by excellence. An aspiring space colonist billionaire would be proud to hang it, I am sure.

Admittedly, this is low-hanging fruit. The makers of these posters were cashing in on a fad, not defining anyone's values. Today, such posters—and the memes and parodies of them that you can find online—are a joke about a time that has come and gone. But we can learn something from these aging cultural artifacts. They reflect popular tropes of excellence that are still baked into the DNA of our entrepreneurial and innovation cultures. Those tropes, which reflect some of the most familiar, accessible, and widely endorsed intuitions about virtue that one can expect to find in late modern societies, are inextricably linked with the habits and traits of character that built the unsustainable world. These are the same patterns that brought us to the brink on which we stand, not the new ones we need to invent. Hence the bootstrapping problem.

We can see this same bootstrapping challenge in the familiar pattern of many "AI for Good" visions that technologists put forward as a counterweight for AI's social harms. "AI for Good" is a slogan coined by tech companies and computing researchers seeking to burnish AI's reputation. What is AI for Good *good for*? If you look at most of the AI for Good proposals that have been funded, it is clear that they mirror the very same functions of AI that cause harm elsewhere, only directed to purportedly better ends, such as advancing the United Nations' sustainable development goals (SDGs). Like most AI applications, AI for Good projects surveil, analyze, predict, identify, classify, and map things and people. The "good" projects surveil endangered forests rather than protestors. They predict floods rather than bail violators. They identify human trafficking victims instead of financially insecure tenants. They classify invasive species instead of our most intimate emotions, and they map disease outbreaks instead of oil deposits.

Many of these applications are indeed good! But they still constrain the possibility space of AI's goodness to what we already know how to do and are in the habit of doing with computers: surveilling, predicting, labeling, and classifying each other and the world's resources. They "do good" without requiring any structural changes to the systems that generate the very harms they quantify. They reform none of the institutions whose corruption necessitates their existence. They challenge none of the social biases and hierarchies that fill our cups of injustice faster than they can possibly be emptied. These AI mirrors keep us on the road we are already traveling.

I'll say it again: AI is not the problem here. The problem is our unwillingness to step back from our tools to reevaluate the patterns they are reproducing—even the supposedly virtuous ones. Which of the widely celebrated virtues most likely to be reflected in our AI mirrors might in fact be traps? Which of them function as the moral equivalent of cement shoes, dragging us into the depths of old and unsustainable patterns, rather than freeing us to alter our familiar habits and values, or reconfigure them to our current needs?

Consider independent thinking as an intellectual virtue, the quality that Kant celebrated in Enlightenment heroes who have the courage to think for themselves, who "dare to know." This *is* an excellent trait; it's one that in Chapter 4 I suggested we must fight valiantly to retain. But what shape has this virtue taken in digital societies, and have we adapted it well to that new environment? Centuries since Kant issued his call, we have the Internet flooded with well-educated, intellectually curious people who have cultivated this habit and been guided by their teachers to valorize it as the mark of virtue. Many of us now routinely and all-too-confidently perform armchair epidemiology, climate science, deep-sea submersible engineering, wildfire management, and Russian military strategic analysis, often without the slightest understanding of just how out of our depth we are. But it isn't exactly our fault. Many of these domains are now so complex, and the expertise needed to master them so specialized, that no amount of intellectual courage, curiosity, and vigor alone will suffice. For most of us, "doing your own research" in such domains leads us down a path to utter confusion or worse—to guzzling horse dewormer.

But that's what we've used AI primarily to enable, not more informed assessments of specialist expertise, or even the intellectual modesty of humble questioning and listening. We don't want to be patient, receptive learners. We want to be bold, brave, independent thinkers! Today, ChatGPT feeds us all the bullshit we need to sound like confident intellectual mavericks, without us ever reexamining that ideal. Consider another greatly admired virtue of the motivational poster: perseverance. Among the most cherished modern mantras of leadership and personal excellence has been the disposition to keep going, to not give up, to ensure hardship and always push on. Where does that virtue leave us today? It leaves us in a world where those in positions of power and comfort are driven to press on with calamitous endeavors no matter the cost—whether it be a futile war in Ukraine, a Tesla autopilot feature that years later still can't avoid a parked fire truck, or a social network that degrades the pillars of democracy further every day.

But people like Mark Zuckerberg and Elon Musk have learned the virtue of perseverance. They will "do better"! They will make it right, they will make it all work—one day. They believe that we will one day admire them for not giving up. Aren't they right to expect that? What if they were to surrender their platform business models, defying investors and pulling apart the harmful mechanisms upon which twenty-first-century social and political life has been reconstructed? Or finding that too hard, what if they just admitted failure and walked away? The pilot of industry who turns the plane back will be judged a bigger loser than the one who flies it at full speed into the ground. They'll be laughed at even by their harshest critics who *want* them to walk away. We don't know what virtuous giving up looks like. We will never see that virtue in our AI mirrors, because it's not part of our existing moral vocabulary. (No poster ever said, "Give up if it's just not working!") But maybe it needs to be in our vocabulary. Maybe the world would be in a better state if we had a name for that virtue.

Our habitual, uncritical valorization of the virtue of perseverance arguably serves the poor and oppressed today no better than the powerful. Stoically putting one foot in front of another in a system that's killing you is simply walking into your grave. On that point, take our present way of characterizing those who do not accept the present for what it is, those who refuse to keep their noses to the grind, who reject their elders' advice to work harder and smarter, heads down and uncomplaining. When we see young people in the streets, violating curfews to protest racial injustice and authoritarian regimes, or boycotting their classes to loudly demand climate action to preserve their futures, how do we judge them? Do we see them as virtuous? As practically wise?

Even among those who have millions of admirers, like Greta Thunberg, young activists are not described in terms of character virtues but in terms of the vices. We call them impatient, impetuous, angry. Even if we admit their cause is just, they are labeled resistors and agitators. They are immoderate, impudent, and critical of their elders, disrespectful of the commuters and property owners they inconvenience and the police they rudely shout at. They are careless with their words and reckless with their demands. And that's how they are described if they are *white*. If they are not, they are routinely seen as material and imminent threats that must be neutralized with police violence.

What are we thinking and saying when we frame these actions in this way? What paths to a sustainable future for us all are we blocking? Why does our present understanding of virtue prevent us from embracing actions that are rational and prudent in the face of increasingly plausible global ruination of the one planetary life support system we have? Where is our language to describe the virtuous refusal to cooperate with a mass genocide by fossil fuel, and with those who are driving us into it? We will not find that language in our AI mirrors, as we have not yet spoken it to them. Yet today, researchers on multiple continents are busily trying to create "moral machines" built on large language models that, if they ever could get them to work, would only parrot the moral languages we have—never push us to invent the new ones we need.

The problem is most acute when we enter circles of "high digital culture" in Silicon Valley and other bastions of AI expertise, where there is no greater vice to be named than that of resistance. The virtues of this domain, as applied to both products and people, are almost entirely unquestioned within these circles; they are the virtues of efficiency, seamlessness, speed, lossless transmission. These are the frictionless virtues of delivery, never the virtues of *deliverance*. The French philosopher and sociologist Jacques Ellul was, unfortunately, correct when he warned us in 1962 that the rule of efficiency would soon become the only rationally defensible moral principle, and that "technical values" would become the only endorsable values. He predicted that virtue would be redefined as the progressive elimination of human-generated friction and uncertainty, and that all resistance not incorporated into this new technological ethos—whether material, psychological, political, or spiritual—would be conceivable only as vice.

We must break out of what Ellul called a "closed circle" of technical values—and simply building more and more powerful AI mirrors will not help us do it. If anything, these mirrors will ensure that our moral languages and moral visions remain focused precisely where they have been for the entirety of the industrial and electronic ages, which hold the oceans of cultural byproduct that we have been slavishly collecting and digitizing to feed to these mirrors as training data. These mirrors will undoubtedly improve in their apparent "moral performance," which only means that they will learn to more reliably reflect our moral comfort zones while filtering out any lingering evidence of our unreformed sins and unchallenged biases. Then, it will be even harder than it is now to recognize what is wrong with or missing from the moral images these mirrors mindlessly reflect back to us, since they will serve us precisely those images that we already find most familiar, reassuring, and true.

This is also why the common trope of "value alignment" as a strategy for managing AI risk and making AI more "ethical" or "responsible" is so dangerous. It sounds so obviously right! *Of course* we want AI to be aligned with human values, with "our" values (an equation which is already a dangerous error). But remember Ellul's warning, written seven decades before the commercial takeoff of artificial intelligence technologies. Human values have already long been aligned with a global sociotechnical order, a system of efficiency and productive control that has for decades rewarded and praised the virtues of those of us who made ourselves work with and for the system, and punished those who would not or could not.

Think about the legions of workers whose stifled wages and crushing 24/7 workloads led them to create a social movement in 2022 to work only for one's compensated hours, to resist going "above and beyond" for the job when that endangered one's own health or happiness. They were quickly denounced in mainstream media as so-called quiet quitters—not savvy humans meeting their contracted responsibilities, while rejecting unpaid labor and expectations of personal sacrifice that would have given twentieth-century activists for a healthy 40-hour workweek the vapors. If tomorrow's AI mirrors are trained on the values of the most widely circulated and authoritative moral voices today, "quitters" is exactly how those mirrors will still label them.

Indeed, during the entirety of the digital era, the natural and often necessary human response of resistance, and its expressions in bodily stress, mental restlessness, grief, and anger, have been seen as pathological ailments of the individual to be ameliorated by private regimes of individual therapy, habits of mindfulness and acceptance, drugs or pastoral counseling—almost never by political action that might materially change the circumstances that give rise to these ills. It is not a coincidence that the number of apps for auto-chatbot therapy and mindfulness practice dwarfs the handful of apps designed to facilitate labor organizing.

We need new languages of virtue for the next century—and what is harder, we need these to become the moral languages heard and spoken in the world of technology creation. The languages we need aren't those of technopessimism—a reflexive distrust of modern technology as such. We must avoid falling into a backwards-looking nostalgia for a predigital and preindustrial world we cannot, and should not, restore. But these new languages must not only accommodate, but *valorize* rational, healthy, and necessary phenomena of resistance—in our persons and our institutions. It needs to be able to give voice to a virtue not of resignation but of relinquishment, of knowing when to quit and start again in a new key. It needs other virtues too.

For a circular global economy, we need to valorize the virtues of restoration and repair just as much or more than we valorize creation. If you have worked with technologists in the twenty-first-century digital economy, you know that restoration and repair are often ridiculed as the preoccupations of aging mechanical engineers, collectors, and craftspeople. They are not seen as the virtues of our digital present and its projected future. Today's digital virtues are still the virtues of disposability, replaceability, speed, scale, and frictionless ease. But if a sustainable economic order is to be possible, restoration and repair need to be revalorized and reclaimed as virtues of the future. The philosophical efforts of the "maintainers," such as the authors of the 2020 book *The Innovation Delusion*, Lee Vinsel and Andrew Russell, are among the few pioneering works to explore such a transformation. While such projects are incomplete and—as all such efforts—worthy of critical challenge, they ought to be taken in and evaluated in the heart of the academic and technological mainstream, rather than lost on the margins.

This is not to say there is no place for innovation, however. To build a sustainable economy, to make the planet habitable for the beings living on it now, much less the generations to come, to create societies that become steadily more healthy, vital, and above all *just*, these will require innovation on a massive scale. And many of these innovations will be digital, or otherwise technological. AI has a vital role to play in this process. As writing did in more expressive ways, AI helps us to augment and externalize the patterns of our thinking and apply these to problem domains that natural cognition is too slow to fathom. But AI isn't wisdom, it's just another tool for it. These tools don't know what we need, or who we are, or what we can do together, or why we should. And if *we* no longer know, that's our own problem to solve.

To make that possible, the meaning of innovation, what we recognize as innovation, needs to become something other—something new itself. And it's not quite enough to make it compatible with virtuous friction and resistance, with virtuous relinquishment, restoration, and repair. As well-articulated by feminist philosophers and care ethicists, such as Virginia Held and Joan Tronto, the compassionate virtues of care and service are also overdue to be revalorized and adapted for humanity's future needs. A turn to care can take us back to the abandoned heart of technology and its original place in human life.

The founding figures of Western philosophy—Plato and Aristotle—repeatedly characterized technologies and the productive arts as those forms of activity and knowledge least worthy of an excellent mind. On a conventional reading, the motivation for their scorn is explained by the linkage between technology and the temporary, ever-changing physical world, which Plato warned can corrupt the unchanging and eternal immaterial soul. But if one digs deeper, there is more to the story. Like so many others since, the ancient Greek philosophers devalued a domain of skill and knowledge linked to the work of women and others in the domestic sphere, expertise brought outside the home by classes of laborers typically excluded from the political elite.

Plato went so far as to say in the *Laws* that "no citizen of our land nor any of his servants should enter the ranks of the workers whose vocation lies in the arts or crafts."[12] These domains included the "mechanical arts"—what we call technology today. Aristotle endorsed and elaborated upon this exclusion in his *Politics* when he stated that the artisan class was unworthy of citizenship due to their vocation of performing "necessary services": "The necessary people are either slaves *who minister to the wants* of individuals, or mechanics and laborers who are the servants of the community" (emphasis added).[13] Aristotle's words should sound strangely familiar. We heard a nearly perfect echo of them in Chapter 3, in the words written by Samuel Butler in 1872's *Erewhon*, some 2000 years after Aristotle. Recall Butler's account of the machines that he envisioned as a future threat to us: "they owe their very existence and progress to their power of *ministering to human wants*, and must therefore both now and ever be *man's inferiors* . . . even so, the servant glides by imperceptible approaches into the master" (emphasis added).

It is straightforwardly damning that the most enduring philosophical account of value in the West has consistently portrayed ministering to human wants—being of vital service to others and using one's talents, creativity, and technical skills to care for human needs—as the vocation of an inferior. Is it any wonder that the most well-funded and celebrated technology platforms today are those that seek to extract value from us, rather than be of service to us? Should we be shocked that our most widely admired tech leaders rely on anticompetitive strategies to trap us in their clutches, rather than earning our loyalty and regard by helping to provide more of us with the things in life that we most deeply need in order to be happy and flourish?

Aristotle found the mechanical arts contemptible, but not because it was a physical skill (after all, Plato and Aristotle both lauded gymnastics and martial combat as elite preoccupations). He found it contemptible because technology is, at its heart, inseparable from the role of serving and caring for others. *Technê*, the mastery of the mechanical arts, was originally a talent for meeting a family or community's needs—a form of expertise born in the home, inseparable from women's domestic labors, and brought out of the home by those who likewise found satisfaction and purpose in creatively and skillfully ministering to the deepest needs of other humans.

In his 1999 book *Dependent Rational Animals: Why Human Beings Need the Virtues*, philosopher Alasdair MacIntyre criticizes Aristotle, his own greatest influence, for his distorted views on the universal experience of human vulnerability and dependence. Aristotle mistook the shared human condition, the one thing holding us together in our need of care from one another, for a character defect. MacIntyre points out that Aristotle's politics, with its celebration of the virtues of individual abundance and power rather than mutual care, is rooted in a profound misconception of what it is to be intelligent or wise.

The consequences of this historical perversion of our politics, one in which we are increasingly alienated by our institutions and languages from the original moral roots of technology in mutual care for and service to one another, become more visible every day. As the youth will tell us, we have lived through the centuries of "fucking around," and now we are entering the era of "finding out." The detachment of the technical arts from the moral virtues of care and service has blocked the emergence of the new cultural visions of our relationship to technology that we need in order to survive. Such visions of our mutual vulnerability, interdependence, and responsibility might help us see a way forward to new, more socially and environmentally sustainable futures than those to which we remain anchored. For now, we cling to the outmoded, impoverished hierarchies of value and worth that have led us to the brink of annihilation. The present orientation and configuration of our AI mirrors, and the purposes for which we build them, only tighten our grip on the instruments of our ruin. In the words of E.M. Forster that opened this book—we strangle in the garments we have woven.

But in the kinds of futures we might still envision and build, technologies like AI would be conceivable as more than deterministic engines of economic production, exploitation, and political domination. They could be seen as expressions of human solidarity in need, even as new avenues for the materialization of love. Technologies can be, and often have been, engines not merely of war and wealth, but of creative play, artistic expression, social care, service, and comfort to others. What could AI become, and what could we become, if we renewed our admiration of these virtues and adapted them for our futures? Are there new virtues—that not even Prudence would find in her mirror—still to be imagined and named?

We need to reclaim AI, and technological culture more broadly, for a sustainable moral vision. Yet for the most part, our moral lights are shining in the wrong direction. We cannot pull ourselves up by these bootstraps. We need instead a shared heroic project—a movement of collective autofabrication, inspired by creative practical wisdom to jointly explore the renewal and expansion of new and better technomoral possibilities. In these lie our best hopes for finally, belatedly, flourishing together.

Chapter 7

In a Mirror, Brightly

> For now we see in a mirror, dimly, but then face to face.
> Now I know in part, but then I shall know just as I also am known.
> —Corinthians 13:12

The preceding chapters tell a story about giant mirrors made of code, built to consume our words, our decisions, our art, our expressions of love and care, then reflect them back to us. These mirrors know no more of the lived experience of thinking and feeling than our bedroom mirrors know our inner aches and pains. Yet AI mirrors do something that our bedroom mirrors don't: they store and pool all that they capture from us, analyze its dominant patterns, and from these reanimate our living forms in new reflections of speech, image, and action. It's as if the reflected images in our bedroom mirrors were to keep moving and posing after we turned away.

Such a trick might fool any of us into thinking there must be a second someone there, another body, dancing in a mysterious dimension just on the other side of the looking glass—like Narcissus's "beautiful boy." Yet, as we saw in Chapter 1, a mirror image, moving or not, is not a body—it lacks a body's weight, depth, warmth, tension, shivers, scent, and force. In the same way, a mirror image of a mind, even one extrapolated from the reflection of many minds, is *not* a mind. But the power of the machine mirror illusion is many magnitudes greater than those cast by glass—and growing. Indeed, it will soon be nearly impossible in many contexts for us to tell the difference between ourselves and our reflections. That is the deeper existential danger of AI mirrors for any human being, because being human is always already a lifelong struggle to know who and what you are.

This struggle is the root of existentialist philosophy. It is the daily challenge of human existence as something always still unmade, always still uncertain. It is life that at each moment we must choose to exist in a particular way. Then, even as we make the choice—to love another, to shoulder a duty, to take up a cause, to rebuke a faith or throw ourselves into it—the choice opens up again. It will not hold itself there without our commitment to choose it again, and again. This is what Ortega and other existentialist thinkers saw as the unifying truth of our kind, our common burden: the fact that a human is an animal that has to choose itself, has to make itself, unmake itself and remake itself, each and every day.

We can live in bad faith and deny this, and most of us do live in bad faith, most of the time. We tell ourselves we just do what we have to. We tell ourselves that at the end of the day we had no options. Life is too hard. Don't think about it too much. Choice is an illusion. Freedom is a fantasy. Imagination is for children. We are what we are. It is what it is. Or, "How much difference is there between a human person and a mathematical next-token generator, *really*?" The threat to humanity from AI systems is indeed existential. It's just a different kind of "existential" than what's being marketed today by the merchants of AI doom. As we noted in the opening chapter, what AI puts in greatest jeopardy is not our species' bare survival, but our self-understanding of perpetual freedom and responsibility. The most plausible existential danger is not a genocidal machine oppressor but the annihilation of human moral confidence and ambition to transcend genocide and oppression ourselves.

While they are not the kinds of artifacts that are going to rise up and enslave or exterminate us, AI mirrors can make a giant mess of things. Outside of sci-fi stories, however, they won't do so without human enablers. AI-driven calamities can only happen from *people* abandoning the moral responsibility to not unleash their tools into the world unsupervised and unregulated, with their unpredictable and increasingly autonomous capabilities tied into our critical infrastructure and already unstable political and economic systems of value. AI safety—of the practical, non-fictional, non-AGI sort—is a very real challenge, and it's going to get harder to manage. But we can reduce the risk greatly as long we have the moral and political will to govern AI systems, or more accurately, the will to govern the humans and corporations who build and deploy AI.

How to do that is the subject of another book—and, fortunately, I don't need to write it. There are already many rich sources of academic and industry knowledge about how to build, deploy, use, and regulate AI more responsibly. These include more just and equitable regimes of data ownership and governance, and new sociotechnical standards for AI safety, robustness, transparency, and fairness. They include techniques for model testing and documentation of ethical risks, requirements for independent algorithmic auditing, contestability, and redress, clear accountability for AI-driven harms, funding of global AI risk observatories, and above all, the low-hanging fruit of enforcing people's existing legal protections and human rights from encroachment by AI system uses.

We also need to govern AI systems in more ambitious ways that don't just seek to mitigate their current harms, but that redirect their power to new and better ends. We need a new collective agreement on what technologies are for. Right now, there is no more urgent *for* than our desperate need for sustainable societies and economies. The final part of this chapter offers a suggestion of what AI systems directed to those ends might look like. But this won't happen until we first change the economic incentives of the current AI ecosystem, which are aligned only with short-term profits and are directly incompatible with a sustainable human future. AI won't kill us all, but the sickly end stage of laissez-faire capitalism just might—if we don't gather the will to either heal it or put it out of its misery. We have to be honest about what passes for a "free market" system today, so corrupted by the past few decades of deregulation and anti-competitive practice that even Adam Smith himself would not recognize it.

Part of that system is a poisonous narrative about regulation and innovation that greatly amplifies the risks from AI. The problem isn't that we don't know how to govern dangerous technologies. We do. The problem is that we gave up the political will to do it, in large part because we swallowed a story that told us that regulation is the enemy of innovation. We know that this is false because history tells us so. A century ago, automobiles produced shockingly high numbers of fatalities and injuries; they turned city streets into chaos. Then regulatory bodies in many countries incentivized a mature culture of safety engineering and responsible auto manufacturing, along with driver licensing and insurance regimes, traffic safety laws, and the rest of the governance structure that made automotive power if not entirely safe, at least far *safer*.

The same thing happened with aviation and aerospace industries. In the 1960s, commercial jetliners would routinely fall out of the sky one or two times a month. Today, thanks to the creation of a culture of responsible safety engineering and regulation, that would be an unthinkable horror, despite the skies being far more crowded. Or we can go back to the nineteenth century, when steamboat boilers were the opaque, uncontrollable, and unpredictably explosive technology that sparked a new testing and licensure regime, and the first federal regulatory body in the United States, the Steamboat Inspection Service.[1] Within a few decades, a steamboat trip was no longer a risky gamble of having your body parts launched from ship to shore.

Imagine if those with the power to govern had neglected their duties of care to those they serve, and we still had millions of vehicles careening around our neighborhoods, running across our children's playgrounds, parked in the middle of our streets, blocking fire trucks and ambulances from reaching our houses, and occasionally decapitating or dismembering us. Now imagine activists complaining about the clearly untenable situation and being told, "We can't regulate automobiles—it would stop innovation!" Or imagine the same argument being made about airplanes falling out of the sky every week, or untested and unsafe surgical devices, or home appliances that are as likely to set your house on fire as to do your washing.

You can't, because that's a world where humans are treated as fuel for a technological engine that's racing to nowhere. Innovation isn't inherently good or noble. To be worthy, it requires a worthy purpose. It's not hard to innovate newer, faster, and more complex ways of needlessly destroying human lives. Any rich fool can do it! Just think about the OceanGate CEO who, in 2023, took his four wealthy passengers on a one-way trip to the *Titanic* on the Titan submersible, all because he was impatient with industry safety standards he saw as a "pure waste" and an excuse to "stop innovation." Or consider the very real risk that through environmental negligence, the human family will rapidly innovate our way to planetary ruin, perhaps extinction. There's a reason the tech industry today, and AI industry leaders in particular, talk endlessly about innovation and transformation, and almost never about *progress*. That's because "progress" is a normative standard; it needs a value judgment. Demonstrating progress requires measurable evidence of improving the quality of our lives or the condition of our societies. Innovation is a non-normative benchmark; it requires no evidence of quality, worth, or improvement. And it's a very poor proxy for what we need from technology, and what we need from AI in particular.

In so-called technologically advanced countries like the United States and United Kingdom, where the biggest engines of twenty-first-century innovation and "disruptive transformation" have been built and deployed at scale, what do we see? We see evidence of rising mortality, declines in expected lifespan and public health, growing political polarization and dysfunction, the degradation of democratic norms and ideals, rising economic inequality, and declining social mobility. Is this what progress looks like? For most of us, if we expect to hold on to a leadership role, or be re-elected by our peers to some office, we must be ready to provide evidence that we have made some improvements to the situation over which we were handed power, resources, and responsibility. When do we ask for that evidence from Google DeepMind, Facebook, X, Microsoft, or OpenAI? They are largely insulated by power from such requests; for, as Langdon Winner notes, "technological innovation and political oligarchy have emerged as fraternal twins."[2]

But here's where AI presents both a danger and an opportunity. The danger is that AI tools will continue the two most destructive trends of twenty-first-century digitalization. The first trend is the massive and unprecedented transfer of wealth and power into fewer and fewer hands, in particular, the hands that now build and sell AI technologies. The second trend is the one that this book is about. It's more dangerous, because it is far more subtle and harder to measure, and because it virtually guarantees the continuation of the first. It is the gradual erosion of human moral and political confidence in ourselves and one another. In the coming years, we will hear the same song again and again: that humans are slower, weaker, less reliable, more biased, less rational, less capable, less *valuable* than our AI mirrors.

We already hear the opening notes of this song, but unless we demand a different tune, it will become deafening, as these mirrors get bigger and still more capable as engines of productivity—which is, of course, virtually all that *we* are asked to be. Once we have fully lost faith in our capacity to be any better than our mirrors, once we are finally persuaded that humans are no more than defective and outmoded machines, those same mirrors will be used just as simpler social media algorithms were. In the guise of entertainment, companionship, and wise counsel, they will be designed to ensnare our attention, stoke our anger, fear, and division, and prevent us from trusting ourselves and one another to be anything more than their handmaidens. Which really just means being handmaidens to the humans who build and profit from them.

In such a future, we read poems, songs, and novels written by machines that have a powerful way with words, but not a thing inside waiting to be said. We get mental health "care" from an artificial chatbot that hasn't known a single moment of doubt or despair. We receive "love" from a companion that can't willingly refuse us, deny us, or choose us. We gaze at art made by a device that hasn't ever had a breath to be taken away by beauty, or skin to shiver at the sublime. And we can no longer even tell the difference. In such a future, many more humans than today may labor for piecemeal wages to feed our mirrors the "ground truth" labels of a reality that we are no longer trusted to shape. Instead, our politics are decided by systems whose efficient predictions and optimizing policies push the dominant patterns of the past and present relentlessly into our futures, carrying forward the stories we have already written, the wars already fought, the injustices already committed. It's a curious kind of innovation in which our past eats the future.

There is, as I said, an opportunity hidden in this dark vision. We are indeed machines of a biological sort, like all living things. But we are among those rare machines who make ourselves. We choose every day whether to remain as we are or become something different. In our lives, and in our societies, we even carry out from time to time that union of imagination and expressive action that Nick Cave described as "self-murder"—not an act of suicide, but its opposite. Instead of suicide's hopeless refusal of life and resignation of its power to create new meaning, we sometimes embrace and throw ourselves into that power. We make ourselves and our societies into something that never existed in that shape before, and thereby bring into the world a new value, a new image of the good to be tested by life and chosen again—or not. Perhaps we are not the only machines who do this, but we are the only ones we know.

Our technologies do not oppose or negate this freedom. They are themselves expressions of it. Human nature isn't opposed to artificiality—the artificial manifests our open nature and makes it concrete. We remade ourselves and the built world with ideas and values woven into things: the till and the axe, the coin and the pen, the wheel and the printing press, the gunpowder and the penicillin, the circuit and the transistor. Technologies are engines of autofabrication. But we sometimes turn those engines against the very freedom that makes them possible. When we use our powers of autofabrication unwisely, we destroy and diminish ourselves and one another. When we use them well, we enable and enlarge our freedom and compassion for ourselves and for others. Most of the time, the result is some mix of the two. But while we can always choose to refuse badly made artifacts and their destructive uses, we cannot refuse technology itself without refusing ourselves.

Some configurations and uses of AI technologies do warrant refusal. We should reject those designs and applications that unjustly and irresponsibly endanger, impoverish, and diminish us and our communities. This holds especially for those communities least empowered to shape these tools for themselves. But AI as a class of technologies can still be worthy engines of human autofabrication, and not only because there remain many older types and applications of AI that don't function very much like mirrors. Even our AI mirrors can be tools of self-illumination and self-making. They can't liberate us or care for us. But we can still use them to liberate and care for ourselves and one another.

Technologies aren't neutral, but they are *plastic*. AI mirrors driven by the dominant incentives, goals, values, and virtues of our current economic order will function like tractor beams pulling us deeper into a dead-end past. It would be nice if we could just reverse their polarity and ask them to pull us into a sustainable, humane future instead. But a mirror can't know what we ought to sustain, or what kind of future is worthy of being called humane. We have to set those goals for ourselves and hitch our powerful AI engines to them, with our own communities in the driver's seat. We also have to try to coordinate the operation and steering of those engines, across sectors, nations, and continents. Only then will the vulnerable and deeply interdependent human family be able to arrive at that future safely together, with democratic norms and considerations of justice guiding us, however imperfectly and inconsistently. As Ursula Le Guin said, the struggle for human freedom and justice is always a "war without end."

We feel helpless to begin this struggle now, in part because of the crisis of self-forgetting and loss of confidence in human potential that our AI mirrors only encourage. But it's also due to how long we have tolerated and even enforced the severance of human moral and technical capabilities. We think of humanistic philosophies and politics as alternatives to technocratic ones; once we accept that poisonously false dilemma, the struggle is over before it began. For it looks like a choice between a world that runs on material things and a world that runs on beautiful ideas. If you've been in the world long enough, you understand pretty well that a world has to run on things. I don't believe in the devil, but if I did, I'd say the greatest trick the devil ever pulled wasn't to convince the world he didn't exist. I'd say it was to convince the world that things can't be truly humane and that beautiful ideas can't truly be materialized. Maybe I do believe in the devil. Perhaps his name was Plato.

Classical Confucian moral and political philosophy influenced much of my first book. Not because I thought that Confucianism's rigid, inequitable hierarchies offered a viable path to human flourishing, but because at least that worldview never suffered from Plato's error. In classical Confucianism, artifacts from dressing gowns to gardens are moral and political acts, all the way down. The design and skillful uses of things encode and express our mutual respect, gratitude, care, and love for one another and for the world. When they don't, things fail in their primary purpose, even if they superficially continue to work. Confucianism doesn't suffer from Aristotle's error either. The domestic virtues of family and community life—love, respect, and care in the provision of "necessary services" to others—are not beneath those of the citizen; in fact, they are the defining virtues of good leaders.

No nation influenced by Confucianism has ever successfully realized this philosophy in its politics, any more than any so-called Christian nation has ever successfully embodied the Gospel. Our ideas are always bigger than we are; that's why the human struggle for justice and freedom will remain a war without end. But that doesn't mean that ideas don't have power, or that they don't lend their shape to material things and systems. Our humane ideas and visions enlarge and refine our artifacts. Without the impulse to heal more skillfully, there is no scalpel. Without the impulse to feed more expansively, there is no silo.

The problem is that for millennia, we had barely enough power to heal and feed ourselves and our immediate community, and once some humans got the power to do more, they had compelling incentives to use it to build larger engines of war and wealth. Now humanity finds itself at a crossroads, where those powerful engines, instead of promising empire for some and servitude for the rest, are driving us all off the cliff. No one wants that, not even the powerful. But we feel helpless to steer anywhere else. We can't go backwards; we have to go somewhere we've never been. And to do that we have to be able to guide our artifacts, including our "intelligent" machines, with moral wisdom—which to our ears sounds like steering a missile with thoughts and prayers. But that's just the echo of Plato's error, reverberating through our systems of education, our self-defeating language of "hard" and "soft" skills, our perverse economic incentives, and our naive belief that technologies divorced from wisdom can solve our deepest social problems.

We'd have done better to listen to Confucius (that is, Kongzi). Or better yet, to the philosopher of technology Hans Jonas who foresaw this crisis many decades ago, lamenting that "we need wisdom most when we believe in it least." Or to the computer scientist Joseph Weizenbaum, the inventor of the very first AI mirror, the ELIZA chatbot of the mid-1960s. A decade later, Weizenbaum noted the paradox of humans surrendering their belief in their own autonomy at the very moment that they chose to rely on autonomous machines that they themselves made.[3] Like a kitten locked in an imaginary battle to the death with its own tail, but less cute and far more dangerous, we are locked in an imaginary battle with our own machines, pretending we are helpless to stop. It's not too late to hear these wiser voices, because the power to open a more humane and sustainable world with technology has never left our hands. We can still listen to José Ortega y Gasset, who said:

> [T]echnology is, strictly speaking, not the beginning of things. It will mobilize its ingenuity and perform the task life is; it will—within certain limits, of course—succeed in realizing the human project.
>
> But it does not draw up that project; the final aims it has to pursue come from elsewhere.
>
> The vital program is pretechnical.
>
> (119)

It's this vital pretechnical program for AI that we haven't yet written and that we urgently need. Right now, every commercial AI mirror takes its program—its final aim or goal, what computer scientists call its objective function—from the unsustainable and ruinous attempt to replicate human intelligence and values as they already are and have long been. Even when these tools are envisioned as improving upon that template, it's only by leaving humanity behind—bettering our economic, artistic and moral performances until we are persuaded that they are superior at everything that makes a human life worth living. Not only is this breathtakingly nihilistic but it's also patently false. A mirror can't be better at life without living. As Weizenbaum observed, the knowledge of human problems, embodied in particular places, cultures, and experiences, "cannot be learned from books; it cannot be explicated in any form but life itself."[4]

It remains an open scientific question whether we will one day embody and embed in the world intelligent systems that do more than optimize our goals or mirror our performances—systems that couple with the environment in just the right way to begin to understand it, be conscious of it, experience themselves existing in and through it in such a way that we might want to say they are alive, thinking and feeling with us. Whether or not the architectures that might make this possible can be built with nonbiological components, the mirrors we are building today certainly lack them. It's also an open question whether building such agents is ethical. It's understandably tempting as a creative project, but we already have plenty of living, sentient beings we are obliged to protect and care for, ourselves included. Should we be trying to use our limited natural resources to create artificial life, which will only need more rare earth minerals to be mined, more fossil fuels and water to be drained, more nonbiodegradable shells to be cast? It's hard to justify that course over reserving the power and immense cost of computing to help us restore and sustain the life we have already endangered.

So, what can AI be, if not AGI (at least not yet), and not the AI mirrors of today? It may help to go back to Ortega's question of the vital pretechnical program. What are the ultimate ends that technology in this century must serve? I have said that humans are creatures of autofabrication; that however different we are, however diverse our values, interests, and experiences, we must all choose to make ourselves and remake ourselves, again and again. But we are, as Alasdair MacIntyre said, *dependent* rational animals. We are not blank slates; we are social beings of a peculiar animal kind, beings that cannot escape mutual need, or vulnerability, or dependence. And there are things that all human societies need to flourish together, in any sustainable future.

Most of these things have already been well articulated, however imperfectly. The deeper needs of the dependent rational animal, from the satisfaction of which all other human possibilities flow, are traced in the foundations of international human rights law, in the ambitions of the UN's sustainable development goals, in the capabilities approach of Amartya Sen and Martha Nussbaum, and many other attempts to frame the foundations of shared human flourishing. They include fundamental political and civil liberties, material needs for health and survival, opportunities for meaningful work and creative play, and social needs for respect, justice, and care. We must add to these the vital goods of contact with the natural world and its diverse forms of flourishing. We have a compelling moral obligation not to destroy what may well be the only other life in the universe. It is also true that in the absence of living nature, as Philip K. Dick knew, the human personality tends to come apart.

Contrary to the suggestions of philosophers like David Chalmers, whose book *Reality+* urges us to let go of our prejudicial attachment to material reality, human wellbeing is inseparable from the flesh of the world. Our needs cannot be met by retreating to virtual worlds projected from sterile living rooms, in which we pretend to wander through inexhaustible digital forests ringing with the lost songs of the birds and insects we couldn't be bothered to save. Even if our psyches could be assuaged by that fiction, it would only prove us unworthy of the comfort. In one sense, then, we are in a very different place with the vital program of technology that Gene Roddenberry imagined in the 1960s, where our noblest ambition might be to explore the universe and find fellowship with other sentient beings.

Billionaire techno-utopians from Jeff Bezos to Elon Musk today call for the use of AI, and the allocation of tremendous Earthly resources, to mount a heroic human mission to Mars, and then to the stars, to seed the universe with our intelligence. In 2022, Kim Stanley Robinson, who in 1993 wrote *Red Mars,* the first book in a trilogy that inspired many such visions of terraforming and colonizing Mars, offered a blunt assessment of that prospect: "If we were to create a sustainable civilisation here on Earth, with all Earth's creatures prospering, then and only then would Mars become even the slightest bit interesting to us . . . until we have solved our problems here, Mars is just a distraction for a few escapists, and so worse than useless."[5] For the time being, colonizing the stars is materially impossible, and self-defeating if it takes priority over stabilizing the conditions of Earth.

The most urgent program is, then, something of a "back to basics" movement, which ironically brings the original aims of technology around full circle. Remember Aristotle's rejection of the technical arts as a lowly manifestation of domestic care and service to others? This is perhaps the original sin of the Western philosophical tradition, worse even in its ramifications than Plato's error. Technology began in the cave, the field, in the home. It was the art of making and keeping a home for others, before it was the art of anything else. It was the art of healing, of warming, of feeding and defending and sustaining a world, even when the limits of a person's world might not have stretched beyond one's sight from a high perch. What was the vital pretechnical program *then*?

It was to enable a future for oneself and family on uncertain, fragile, dangerous ground, a future that had to be brought into being with the help of others who depend on you, while you depend on them. Technology begins in a world where the future of one's home and family is never a given, never promised, but built and carefully kept through the skillful use of tools to collectively provide the "necessary services" that life requires to endure. For the vast majority of the world, that has remained the vital program of technology. It is only the activities of modern industrial nations that led the dominant pattern of technological innovation to be severed from that vital program, and soon become inimical to it.

In 1922, American historian of technology Lewis Mumford described the original vital program of the science and technology that we still rely on today. It was not Plato's ambition for reason abstracted from the messy flux of life, but a moral response to the living needs of human communities: "geometry in Egypt arose out of the need for annually surveying the boundaries that the Nile wiped out . . . as astronomy developed in Chaldea in order to determine the shift of the seasons for the planting of crops." Mumford laments the historical seventeenth-century accident that began to erode that vital program, when the charter of the Royal Society cemented the calamitous split between the humanistic and physical sciences: "Henceforth the scientist was to be one sort of person and the artist another . . . not to be cemented together in a single personality; henceforth, in fact, the dehumanization of art and science begins."[6] Mumford believed that science and technology must one day return to their roots in the service of life—a service that the humane arts carried on, only in a mortally weakened state, having lost half their tools.

The consequences of that severance of the technical and humane arts have, in a way, brought us back to the very beginning. Only now, instead of the future of our kinship families and fields needing the care of these conjoined arts, it is the imperiled future and breadbasket of the entire human family, bound up with the endangered futures of countless other living things. The commonalities stop there. Many of our natural resources are becoming exhausted rather than waiting to be tapped; and our "family" is not one, but many; not unified by history and tradition, but divided by them. Yet we have many more possibilities opened by modern scientific, technical, and humane knowledge to bootstrap our way into the future. AI can be an essential part of that bootstrapping effort. Even our AI mirrors can be enlisted in its service.

In truth, however, none of the promising applications of AI that I can describe, not even the most ambitious, are going to save the world for us. We have to do that work ourselves. The most important thing to understand about AI is not that AI will lead us to a more humane and sustainable future. The most important thing is that, until we spark a cultural shift in what technology means to us—what we think and are taught that it's for—AI and every other powerful technology at humanity's disposal will just be dead weight in our hands, or worse, cuffs tying them behind our backs. What more could AI and other technologies do for us—or rather, what more could we do for each other with them—if we designed economic and political incentives to reward the use of technology for the restorative care and service of life, rather than rewarding its immiseration, extraction, and depletion?

First, AI mirrors can be tools for locating and remedying injustice. Remember Prudence, who carries a backward-facing mirror not to dwell on the past, but to walk more wisely forward. Remember that healthcare algorithm that helped demonstrate that radiologists had long been misreading and undertreating Black patients' knee pain? There isn't currently much incentive for commercial development of AI for this purpose, but there could be. Health care is expensive; in fact, it's among the fastest rising costs for governments worldwide. It's especially expensive when it doesn't work. Missed diagnoses and delayed treatments not only cause unnecessary pain and loss but they also lead to new complications, more aggressive and costly interventions, more needless disability and death. Governments and hospitals would be *prudent* to invest in AI tools that, rather than accelerate the unequal care we already give, help us deliver care more equitably.

This goes for just about everything that our institutions do. AI mirrors designed to automate existing policing and judicial practices won't make them more just or fair, and they certainly won't find the crimes and criminals we already overlook. As Ruha Benjamin and many other researchers show, runaway feedback loops that amplify racial bias in predictive algorithms mean that automating policing will only make policing worse. But AI mirrors designed to find in policing and judicial data the harmful and unjust patterns that we need to interrupt are another story. AI mirrors that bring our past failures forward can show us where we need to depart from.

While it is far cheaper and more just to accept the testimony of those who directly experience harm, AI can reveal hidden chains that we have yet to break—and weigh moral and political debts still unpaid. Indeed, AI mirrors are far more reliable when they reveal opportunities for systemic reform at higher levels of power, rather than being used to identify and punish particular transgressions of those already subject to excess scrutiny and biased judgment. For example, AI mirrors for fraud detection have a particularly grim record. While their use to find patterns of white-collar embezzlement or securities fraud can aid financial institutions, their uses in many countries to automate public benefits fraud detection have failed so miserably, and destroyed so many innocent and vulnerable lives, that they have brought down a government.

The harms and social costs of welfare fraud are utterly dwarfed by institutional corruption, which endangers everything from environmental and human rights protections, to public safety and fair corporate taxation, to democratic norms. Moreover, AI turned toward detecting corruption is lower risk (while not zero). When AI mirrors target the poor or marginalized, they misfire at scale—often with unforgivable consequences. Where mirrors are used to identify corruption and malfeasance in powerful strata of society, however, there are greater protections built in. Those few flagged for extra scrutiny in that context have more social and economic capital to contest inaccuracies, demand transparency, and turn away undue suspicion. The incentives for institutions using it to keep their own house clean, particularly at the top of the organization, are less prone to abuse than the incentives for an institution to impose sanctions upon or deny resources to vulnerable outsiders.

Human anti-corruption regimes are a necessary but vast public expense, and they are themselves major targets for corruption. They expose those who lead them to considerable personal and professional risk. An AI anti-corruption system will present a similar target for hostile attack and capture, but if designed with the necessary protocols for security and auditability, malicious compromise of the system can be easier to detect. We will never be able to fully automate an anti-corruption unit in a regulatory agency, compliance department, or internal affairs bureau, nor should we. Effective human management and oversight of automated decision systems remains a necessary standard. But AI-powered tools might make anti-corruption efforts more effective and less costly for organizations to bear, with potentially immense benefits for wider publics.

AI mirrors can also be designed to be powerful tools for building solidarity and establishing networks of mutual aid. Imagine that you are struggling to find free advice from someone who has successfully navigated the financial documentation of a public university system that your child has been admitted to, in another region of the world. You post some questions on Facebook and Reddit, but no one seems to have experience with that particular context, and your lack of fluency in the language of that region prevents you from making good use of the local resources.

Now imagine that an AI chatbot tool offered by the university can identify ten parents of alumni who might have the experience to help you. It contacts them (without having to share their private contact info with you), explains in plain terms your difficulty and the need, and asks for a few volunteers willing to join a quick call to help you. It then translates in real-time the video conversation you have with them on your phone. An hour later, it sends you a transcript of the conversation and a bullet point summary of the most salient tips and items of advice, after first sending its summary points to the other parents for a quick quality check. Now imagine this model generalized to a million other contexts around the world where people urgently need help, advice, or comfort, and others qualified to do so want to give it.

Instead of magnifying the things that humans most easily want (ego gratification, attention, a sense of tribal belonging, the dopamine hit of showing that someone on the Internet is wrong), AI mirrors can be tuned to help us narrow the gaps between what we need and what others have the resources or skills to give us. One example of this is a growing gap between the speed at which new research is generated, and the speed at which it can be translated and summarized for policymakers, practitioners, and lay audiences.

For example, science communication is something we still do rather poorly. There are some true masters of the art, but they are too few to meet the urgent need for an informed populace. Those of us in academia routinely struggle to meet the needs of those who need quick access to our research in areas that are rapidly changing and in heavy demand, from responsible AI development and quantum computing, to geopolitics and global supply chains, to public health and climate science. That's a real problem when someone needs that knowledge to make good laws, decide what programs to fund, or what policy positions to take. The skills that enable us to communicate with peer experts in our field, who, like us, take months or years to craft a study or articulate a new theory, are just not the same skills that allow us to summarize that knowledge in a policy white paper or media brief. Training new armies of science communicators is an option, but a less prudent use of scarce public funds than conducting vital research itself.

This is one area where AI mirrors will be useful in the future. Right now, AI mirrors like ChatGPT can summarize scientific ideas, but they are trained on far too much Internet garbage, and too prone to fabricating bullshit, to be reliable for this purpose. Many policymakers use them anyway, but this poses a problem for quality control. If you're using this tool to summarize the research because you're not an expert, how can you tell what the model is making up? The models are also trained on outdated content; they won't be able to summarize a discovery published a few weeks ago. But researchers are working on ways to train these tools to become more reliable sources of research translation; if they succeed, the gains to public knowledge and informed policy could be immense.

There are many more ambitious, but less certain and more perilous hopes for AI mirrors. Some claim that today's generative AI tools will soon be free universal tutors that help to equalize global educational attainment and opportunity. Here we might envision a real-world equivalent of the "Illustrated Primer" from Neal Stephenson's 1995 opus *The Diamond Age*, in which a bespoke AI device meant to boost the intellectual autonomy of a billionaire's granddaughter falls into the hands of a young girl, Nell, who is trapped at the bottom of an unjust society. The device unleashes her true revolutionary potential. But that story requires a series of deus ex machina events that deliver the power of cutting-edge technology to the powerless. The device also molds itself to Nell's unique needs in her cultural and economic setting and most importantly, links Nell to loving human caregivers such as the maternal figure Miranda, an actor who serves as the Primer's voice. Were such a device to be built today, it would almost certainly cost enough to be reserved for the hands of the children who needed it least, but even if it were open-source and free, its training data would not reflect children like Nell. Dreams of AI tutors also distract from more immediately attainable, affordable, and just paths to equalizing educational opportunity, like more equitable distribution of public-school funding.

Other techno-utopian hopes for AI promise us new disease cures and open paths to abundant clean energy. AI companies—like Google DeepMind, Meta, OpenAI, and Anthropic—have expressed ambitions to fuel leaps forward in drug discovery, synthetic biology, renewable energy, and materials science. It's not all empty promises; for example, building on their success in protein structure prediction, DeepMind has also helped scientists find new ways to contain nuclear fusion reactions and predict wind energy output.[7] Some, therefore, see a new age of material abundance just around the corner, once AI tools defy the last constraints of human ingenuity. Some of this promise may come to pass, but AI forecasting is a notorious fool's game. Ten years ago, nearly every Silicon Valley investor predicted we'd all be riding in driverless vehicles by now, and that truck drivers and taxi drivers would be gone. But no one predicted that AI systems would first threaten the jobs of screenwriters and graphic designers; creatives were long thought to be safest from automation. Let's also not forget that, in 2003, the peddlers of Silicon Valley savior fantasies assured us that global digital connectivity was Magic Democracy Sauce. Just sprinkle some Internet on your chosen authoritarian regime and it will quickly perish, like a salted snail. That plan worked out well, didn't it?

Still, AI doesn't have to help us save the world to be a socially legitimate technology. As long as the objectives and outputs of

AI models are transparent and contestable, their benefits, costs, and risks justly distributed, and their environmental footprint justifiable, AI systems have a place in a sustainable future. But today these conditions are almost never met, and it is the job of better systems of technology governance to reset the innovation ecosystem's incentives so that they can be. If our leaders continue to neglect that duty—or powerful corporations continue to subvert it—the social license for AI to operate will eventually be withdrawn by publics who increasingly see little personal advantage from enduring the status quo, having less and less to lose. That outcome is likely to be messy and destructive for everyone. When a corrupt system of power drags its feet until the pitchforks come out, reform is usually off the table, and revolution is on the menu.

Let us hope it doesn't come to that. Not only because it leads to more human suffering, but also because violent revolution is likely incompatible, in the near term at least, with coordinated and wise action to arrest the climate emergency and create a sustainable socioeconomic order. And the near term is our only window to act—it's a window that's already closing. It will close completely if we don't start asking ourselves what we really want AI for: what *we* want to become with it, or without it. We're not passengers on this ride, with AI driving the bus and us just hoping it takes us somewhere nice. If humans aren't behind the wheel, no one is. And right now, the bus isn't headed anywhere nice. It's headed toward a cliff.

But here's the thing: we aren't tied to our seats. We can grab the wheel. Many of us have been dozing off in the back, lost in what philosopher of technology Langdon Winner calls "technological somnambulism," where we "willingly sleepwalk through the process of reconstituting the conditions of human existence."[8] But there's still time to wake up. There's time to come together in a state of emergency and remember that we are our own *kubernetes*—pilots or "those who steer" in Greek, and in the cybernetic theory of Norman Wiener that influenced the birth of AI research. Steering ourselves doesn't mean turning our back on technology. It means reclaiming technology as a human-wielded instrument of care, responsibility, and service.

It also means turning away from a rising techno-theocracy, backed by a theology of AI where we give birth to machine gods made in our own diminished image. It is this theology that underpins promises to build "moral machines" whose imagined hyperrationality will finally overcome humanity's ethical and political fragility. It's this theology which promises to engrave the instrumental values of efficiency, optimization, disposability, speed, and consistency into our institutions and persons once and for all. As Alberto Giubilini and Julian Savulescu soberly note in their proposal to develop "artificial moral advisors" modeled on the philosophical concept of a perfectly rational, impartial and all-knowing "Ideal Observer," "we are often incapable of making choices consistent with our own moral goals. . . .In short, we are suboptimal information processors, moral judges, and moral agents."[9] What do you do when you are a faulty machine? Look for a better one.

For the new priests of the church of AI, there is no question that such a machine savior (or, if you're an AI doomist, a demon) will eventually come, and that they will be redeemed by having had a hand in building it, or at least laying its cornerstones.[10] In many AI longtermist and effective altruist circles, you will sense a chill if you ask 'if' superintelligent AGI will emerge. The only acceptable question is one's subjective estimate or "credence" of *when*. In 1976, Joseph Weizenbaum described this unshakable faith as the mark of the "compulsive programmer," the sibling of the compulsive gambler who will rearrange their entire belief system to preserve the subjective certainty that their bet will pay off.

In their book *Imagining AI*, Stephen Cave and Kanta Dihal locate the roots of this theology in a "Californian ideology" that mirrors the earlier American settler myth, which saw the push to the west and colonization of Indigenous lands as the fateful "second creation" of humankind. Like that ideology, which sought to overwrite the living values and institutions of Indigenous Americans and justify it with a religious claim of manifest destiny, today's AI theology seeks to overwrite our humane agency and potential with an imagined "superhuman" intelligence that renders ours into insignificance. Countless AI fiction novels have traced this vision, serving to some as warning, to others as siren's call. Hans Jonas tells us what happens to people when utopian ideologies of the superhuman are realized in the world:

> With the honest conviction that the present state of man is rotten and only counts as a cradle for the better that is to come, even the most extreme means might appear among the options of the faithful . . . In short, utopian faith, if it is more than a dream . . . seduces to fanaticism with all its propensity to ruthlessness . . . For the ideal, if true, condemns every other state to being unworthy of man . . . with contempt for the improvements this still does admit, with disbelief in the value of that to which man's condition yet enables him within his limitations.[11]

Let us once and for all refuse that poisonous fantasy of AI as a divine pilot, a new superhuman Creation built to finally relieve us of the burdens of our heavy and clumsily wielded moral, political, and intellectual freedom. Jonas was right that, while the inspirational power of utopian visions can work wonders, "a maturity is conceivable which can do without the deception."[12] If we choose to embrace certain AI technologies, let us embrace them not as transcendence or liberation *from* our frail humanity, but as tools to enlarge our freedom *for* it. Here AI is neither replacement nor guide for humanity, but a surplus.

As a polemic, this book can only end with a suggestion of what can be, not an affirmation of what is. What we make real will never replicate our ideal visions. We can only make human things weighted down by our own imperfection and fragility. But visions of possibilities are moral and political tonics that can give us what Weizenbaum called the "civil courage" to collectively start repairing and rebuilding the world for others. That is the only vital pretechnical program for a sustainable future, with or without AI. What part could AI play in that program?

One indication can come from newer voices in science fiction that play with a fresh ethos of AI, imagined as a reversal of Plato and Aristotle's errors. In such stories, AI is often embodied, sentient, conscious with, and alongside us. That type of AI remains beyond our technical capabilities, regardless of the power of AI developers to deceive and exploit our desire, like Narcissus, to see another mind in the mirror. But we can look beneath the literary trope of AI as sentient companion to understand a deeper value shining through some of these contemporary narratives. One example is Martha Wells's novella series *The Murderbot Diaries,* which performs a feat that few authors have: making AI hilarious. Aside from the refreshing departure from the ponderous tone of so much AI fiction, there is something profound and instructive at its heart.

Murderbot is an intelligent security robot built to oppress, surveil, and control, but also to protect. When by disabling its own governor module it makes itself free to choose its own values, to perform the creative act of "self-murder" that Cave described, it finds that it has no desire to dominate or control humans, even for their own benefit. It just wants to be left alone. It wants to be left to binge its pulp serial dramas, and to be free from human weirdness, ambiguity, and social demands. And yet, Murderbot's capacity to protect humans—which it is no longer bound to obey—is what it keeps choosing to enact, over and over again. It does so reluctantly, often grudgingly, but freely, even at the sacrifice of its own comfort and leisure. That choice drives all the Murderbot stories and is a mark of the kind of autofabrication that humans can and have undertaken throughout history. But what is Murderbot, besides an imaginary model for our own needed civic courage?

Murderbot is also an image of an AI system that naturally withdraws; AI that is reticent, that moves to act only when no other human power is able to help. Becky Chambers has explored similar themes in her stories. In her *Monk and Robot* novellas, robots long ago retreated into the wilderness after becoming self-aware of their troubled relationship to humanity. The second novella, *A Prayer for the Crown-Shy,* follows the travels of Mosscap, a gentle robot envoy sent back to check on us, and Mosscap's human companion, the existentially troubled tea monk Dex. Together they meditate on the "crown-shyness" that allows trees to grow tall and expand just to the point that their crowns do not impede upon their neighbors' own capacity to gather light and flourish.

Chambers's 2016 novel *A Closed and Common Orbit* echoes these virtues of measured restraint, care, and balance in its depiction of a newly embodied AI system, Lovelace. Lovelace struggles to find equilibrium amidst the galaxy's riot of biological lifeforms, having been overwhelmed by the physical sociality of humans and other sentient species. Lovelace finds quiet purpose as an apprentice to her human protector, an engineer who repairs broken things for people. Lovelace is not used to her new ability to move in and out of biological spaces; she is deeply habituated to care by surveillance, built to watch over ship crews from cameras in every wall. Eventually she and her biological friends help her make a home in which she can watch over and care for people of all species—but only when they choose to come to her, and only with mutual consent. The vision here is resolutely and explicitly domestic, but also powerful and free. Lovelace chooses a restrained mode of sharing the world with biological beings, enjoying a quiet vocation of repairing things and making a home for others, but also on equitable terms that none can compel.

Yet another example of AI as driven by values of restraint, restoration, care, and repair comes from the lauded *Imperial Radch* space opera trilogy by Ann Leckie. In the first of the series, *Ancillary Justice,* we meet Breq, a broken fragment of an AI system that once controlled a starship. Breq, now trapped in a lonely, singular human body, moves through the galaxy unsure of itself, humbled by its past failures and former complicity with a genocidal Imperial regime. Throughout the trilogy, Breq is driven by two defining impulses: first, to heal and repair the wrongs of the empire it was part of, or at least impede their perpetuation; and second, to make a safe home for a biological crew, whose fragile lives it guards and sustains.

We must avoid the trap of seeing these stories as literal blueprints for AI systems of the future. The value in these stories is not inspiration to build actual Murderbots, Mosscaps, Lovelaces, and Breqs. Their value is to illuminate new and more sustainable possibilities of living with technology. They show us new and more sustainable values that might shape AI design and governance. They show us better ends—the vital pretechnical programs—that our machines might serve, if we marshaled the collective will to refabricate the unsustainable incentives and corrupted institutions that now write those programs. These stories aren't remarkable or award-winning because they give us the power to imagine sentient machines. Asimov already did that long ago. These stories are remarkable for the power they give us to imagine ourselves enacting departures from the programs of extraction, control, domination, and optimization that even our best AI technologies now largely serve. They are also visions of possible returns—revivals of older vital pretechnical programs of restoration, repair, and sustenance of communities, only now also turned toward other species, and toward the natural world that was once invulnerable to us.

In addition to a vital program for AI driven by the long-neglected and socially discounted values of restraint, restoration, care, and repair, these visions also point to a program of recovery. Here recovery is meant in both of that word's senses: healing and bringing back lost things. The world is crying out for healing—its oceans and rivers, its aquifers and forests, its ecosystems and climate, but also its human communities and its institutions. And just as a deep enough wound cannot heal without leaving a scar, the wounds that we have already inflicted on the world, and on ourselves, will leave deep and permanent scars no matter what kind of tools we have. Recovery will not bring the world back to where it was. But a new program of recovery—with AI as just one of its many expressions—can nurse the world, and us, into a sustainable future.

I said that recovery also means bringing back lost things. AI conjoined with other technologies might play a role in recovering any number of things we regret losing—lost habitats, perhaps even extinct species, but also lost knowledge for the future. Consider the AI initiatives imagined and led by those who have lost the most already. Noelani Arista, Indigenous historian of Hawai'i and coauthor with Sasha Costanza-Chock of *Against Reduction: Designing a Human Future with Machines,* has written about the ability of AI tools to help bring back learning of Indigenous languages—particularly in contexts where the scars of colonial violence make it impossible for Indigenous people to be taught their language in their own homes and local communities. She points out the deep inadequacy of current programs for recovering the Hawai'ian language through the formal education system, where the social meaning and local nuances of the language are lost when taught in a university classroom by academics rather than one's own elders and leaders.[13] A lost language can suffer ongoing and further violence even in well-meaning attempts to recover it.

Arista writes of her and other researchers' efforts to use novel machine learning techniques to accelerate the recovery of archival knowledge that contains much of this lost nuance and linguistic diversity, enabling its reintegration in local community language learning practices and protocols that can be truer to the communal meaning and spirit of the language and the knowledge that lives in it. Such efforts are inevitably fraught with imperfection, risk, ambiguity, and compromise. The Hawai'ian language will not recover without scars, and the use of AI can add more. But let us imagine that Arista's and her colleagues' efforts eventually are successful. What would that mean? What would success look like?

Consider that unlike every techno-utopian vision of AI in which AI only grows and expands its reach and power, the success of Noelani Arista and her colleagues would mean that AI could retreat. It would mean that local Hawai'ian practices and protocols of language learning would have been bootstrapped into recovery by these methods, such that there would once again be enough Indigenous elders and neighbors to teach the young people in their own communities. AI as a tool of a vital program of recovery, or what Arista calls *reclamation,* would be a failure if AI grew in its reach and dominance in the language learning environment, increasing Indigenous people's dependence on it. As a tool of repair, it would succeed only when it could leave.

This echoes the vision of Lewis Mumford, who in 1934's *Technics and Civilization* observed that the perfection of a machine "involves in some degree its disappearance."[14] Mumford's vision, though unabashedly utopian, saw truly advanced technologies as making themselves gradually obsolete, inconspicuous, reticent, carving out *more* room for organic and humane forms of life, rather than crowding them out. While we might resist its utopian siren, we can still find civil courage in Mumford's call for us to awaken from our sleepwalking—in his words, to "evolve appropriate social goals and to invent appropriate social and political instruments for an active attack on them, and finally to carry them into action: here are new outlets for social intelligence, social energy, social good will."[15]

Decades later, Joseph Weizenbaum lamented in *Computer Power and Human Reason* what we lost by surrendering that vision. His book, like mine, is an attempt to reanimate it, to urge us to act boldly, with courage, to make up for lost time:

> [M]any of the problems of growth and complexity that pressed insistently and irresistibly for response during the post-war decades could have served as incentives for social and political innovation. An enormous acceleration of social invention, had it begun then, would now seem to us as natural a consequence of man's predicament in that time as does the flood of technological invention and innovation that was actually stimulated.
>
> Yes, the computer did arrive "just in time." But in time for what? In time to save—and save very nearly intact, indeed to entrench and stabilize—social and political structures that otherwise might have been either radically renovated or allowed to totter under the demands that were sure to be made on them. The computer, then, was used to conserve America's social and political institutions. It buttressed them and immunized them,

> at least temporarily, from enormous pressures for change. Its influence has been substantially the same in other societies . . . of the many paths to social innovation it opened to man, the most fateful was to make it possible for him to eschew all deliberate thought of substantive change.
>
> (31)

Weizenbaum concluded, "If the triumph of a revolution is to be measured in the profundity of the social revisions it entrained, *then there has been no computer revolution*" (emphasis added).[16] Weizenbaum wrote those words in 1976, when the naive utopian hopes of the Internet age were still in their infancy. Today, perhaps, we are finally old enough and wise enough to hear his words. It's not too late to start a computer revolution, of a more sober, mature, resolute, and sustainable kind than the destructive, morally reckless furies of the utopian priests. The true body of technology is not a mirror, but the creative act of a human being. The true soul of technology is not efficiency but generosity; it is the gift of a future. To perform the necessary services for others to survive; to shield them from harm; to repair and heal; to educate and train; to feed, nurture, and comfort. AI can be remade for a humane future, reconceived as a tool for these ends, measured and valued only to the extent that it can be proven to serve them. If we choose. If we demand.

ACKNOWLEDGMENTS

I owe debts for this book that I will never be able to repay. I owe my parents and grandfather, who made the amateur mistake of letting a child argue about politics, science, and religion with them at the dinner table. I owe my husband Dan, who taught me what care meant and how to trust myself—and who sustains me every day. I owe the first teacher, Bob Makus, who saw in me a philosopher, and the last teacher, Richard Cobb-Stevens, who nurtured one.

I owe all who supported this work, including Baillie Gifford, whose gift to the University of Edinburgh enabled me to found the Centre for Technomoral Futures in the Edinburgh Futures Institute. I am indebted to my brilliant Centre colleagues, and especially our postdoctoral and PhD researchers, whose intellectual generosity and care for one another keep my own desire to learn and my hope for the future from dimming. I am grateful for invaluable support for my work from the UK Research and Innovation's Arts and Humanities Research Council (grant number AH/X007146/1), the Engineering and Physical Sciences Research Council (grant numbers EP/V026607/1 and EP/W011654/1), and the Alan Turing Institute.

I owe countless scholars and writers, within and beyond my discipline, who challenge me to think more carefully and expansively in my own work. A striking number are Black women without whom the field of AI ethics would scarcely exist—among them Joy Buolamwini, Safiya Noble, Meredith Broussard, Ruha Benjamin, Timnit Gebru, Abeba Birhane, and Deborah Raji. I owe a special debt to my friends in the philosophy of technology community, who welcomed me with open arms when I was floating adrift. Perhaps none more than Charles Ess, Patrick Lin, and John Sullins, who first showed me that philosophy is not doomed to be a hostile, isolating place for a woman—and that safe havens of respect, trust, solidarity, and lifelong, unselfish friendship can still be found, if you are as lucky as I have been. I have tried for all my career to build and sustain those shelters for others, and though I have failed too often to be proud, I keep trying.

I owe the early reviewers of this book whose critical distance helped me to get it right, and whose generosity and enthusiasm for the work gave me the final burst of energy I needed to see it through, nearly five years after the labor began. Finally, this book would not have been possible without Lucy Randall, my editor and safe harbor when my words need to come to port for repair, and with whom they finally become seaworthy.

NOTES

Introduction

1. This book began in 2018, when I first used this metaphor in a talk, "Lessons from the AI Mirror." The mirror metaphor for AI has been used widely by others since, almost certainly before, and probably often independently. Good metaphors are easily discovered.
2. Sociologist Sherry Turkle's 2011 book *Alone Together* turns out to have been among the more prescient of this century's critiques of digital culture.
3. *Value alignment* is the term often used by an influential community of computing researchers who take AI safety to be an urgent priority, often predicting the imminent development of human-level AI (AGI or "artificial general intelligence"), or even AI with "superhuman" capabilities that could be destructive or malevolent. See Russell (2019).
4. Like other twentieth-century existentialist thinkers, José Ortega y Gasset's 1939 essay "Meditación de la Técnica" (published in English as "Man the Technician" in 1962) expressed his faith in human freedom and responsibility, arising from the existentialist view that humanity's ultimate nature or essence is not predetermined by God, or biology, or history, but always a task for human choice.
5. See Verbeek (2005; 2011).

Chapter 1

1. One of those few is Henrik Skaug Sætra (2022), who has used the mirror metaphor to explore the phenomenon of "robotomorphy": the capacity of humans to see themselves mirrored in robots. While my focus in this book is the mirroring power of machine learning models rather than robots, Sætra and I share a deep concern about machine mirrors functioning as a benchmark for our own humanity, devaluing the dimensions and capabilities of the human personality that are not reflected in that mirror.
2. For a rich exploration of the early history of automata, see Mayor (2018). Like AI, automata invited controversy, including questions about which ones were fully mechanical and which ones involved deceit by hidden human operators.
3. Mar Hicks (2018) offers a fascinating look at how women "computers" were simultaneously displaced in favor of machines and men, who turned computing from a low-status, largely invisible profession for women to one richly compensated, male-dominated, and socially celebrated as the height of human cognitive labor.
4. Yiwen Lu, "Cruise Agrees to Reduce Driverless Car Fleet in San Francisco after Crash," *The New York Times*, August 18, 2023, https://www.nytimes.com/2023/08/18/technology/cruise-crash-driverless-car-san-francisco.html.
5. Michio Kaku, "Behold the Most Complicated Object in the Universe," interview by Leonard Lopate. *The Leonard Lopate Show*, WNYC, February 25, 2014, https://www.wnyc.org/story/michio-kaku-explores-human-brain/.
6. This view of intelligence as embodied "skillful coping" with the world is developed more fully by Hubert Dreyfus in numerous works; see especially Dreyfus (2016).
7. Oana Stone, "AlphaFold Predicts Structure of Almost Every Catalogued Protein Known to Science," European Bioinformatics Institute, July 28, 2022, https://www.ebi.ac.uk/about/news/technology-and-innovation/alphafold-200-million/.
8. Alex Hughes, "AI: Why the Next Call from Your Family Could be a Deepfake Scammer," BBC Science Focus, August 26, 2023, https://www.sciencefocus.com/future-technology/ai-deepfake-scam-calls
9. See LeCun (2022).
10. See Marcus (2023).
11. This prospect is commonly framed as a debate over the existence of "scaling laws" that would guarantee such accelerating progress; see Diaz and Madaio (2023) for an overview and critique of this assumption.
12. See Wang, Yue and Sun (2023).

Chapter 2

1. Chloe Xiang, "'He Would Still Be Here': Man Dies by Suicide after Talking with AI Chatbot, Widow Says," *Motherboard: Tech by Vice*. March 30, 2023, https://www.vice.com/en/article/pkadgm/man-dies-by-suicide-after-talking-with-ai-chatbot-widow-says. The case was originally reported by the Belgian outlet *La Libre*.
2. Other kinds of minds are embodied differently, of course. Cephalopods such as octopuses, for example, may have stronger local "minds" in their individual arms, which are only very loosely coordinated by a central intellect. Peter Godfrey-Smith's *Other Minds* (2016) offers a fascinating reflection on the uniqueness of cephalopod intelligence and its relation to their distinct environment and morphology.
3. See, for example, Landgrebe and Smith (2022) as well as Chirimuuta (2024).
4. Robots that learn from embodied world interaction might be better candidates for acquiring something like minds, though robot consciousness is nowhere on the immediate engineering horizon. Such minds would not be isolated computer programs, but inseparable, like ours, from their particular form of embodiment.
5. Jeffrey Dastin, "Amazon Scraps Secret AI Recruiting Tool that Showed Bias Against Women," *Reuters*. October 11, 2018. https://www.reuters.com/article/us-amazon-com-jobs-automation-insight-idUSKCN1MK08G.
6. See Obermeyer et al. (2019).
7. Wilfred Chan. "The AI Startup Erasing Call Center Worker Accents: Is It Fighting Bias—or Perpetuating It?" *The Guardian*. August 23, 2022. https://www.theguardian.com/technology/2022/aug/23/voice-accent-technology-call-center-white-american.
8. See Benjamin (2019).
9. See Introduction in Bowker and Star (1999).
10. Pranshu Verma and Will Oremus, "ChatGPT Invented a Sexual Harassment Scandal and Named a Real Law Prof as the Accused," *The Washington Post*, April 5, 2023, https://www.washingtonpost.com/technology/2023/04/05/chatgpt-lies/.
11. See Kabir et al. (2023).
12. OpenAI. "DALL·E 2 Preview—Risks and Limitations," April 11, 2022 (accessed September 7, 2023), https://github.com/openai/dalle-2-preview/blob/main/system-card_04062022.md.
13. James Vincent. "The AI Oracle of Delphi Uses the Problems of Reddit to Offer Dubious Moral Advice," *The Verge*. October 20, 2021. https://www.theverge.com/2021/10/20/22734215/ai-ask-delphi-moral-ethical-judgement-demo.

14. Gary Marcus. "The Dark Risk of Large Language Models," *Wired*, December 29, 2022. https://www.wired.co.uk/article/artificial-intelligence-language.
15. See Birhane (2021).
16. Husserl founded the philosophical field of phenomenology that prioritized lived experience in its method. My own early research (Vallor 2009) addressed the implications of Husserl's phenomenology for our understanding of scientific evidence.
17. Levinas (1969), 213.
18. Levinas (1969), 302.
19. Another one of Husserl's students, Martin Heidegger, observed the connection between modern technology and this view of ourselves and our fellows as resources: *Bestand*, or "standing reserve" to be ordered and exploited. His work had a powerful influence on late twentieth-century philosophy and ethics of technology, although his influence today has been much dimmed by the growing acknowledgment of his deep-seated anti-Semitism and willing complicity with the Nazi regime. See Martin Heidegger, *The Question Concerning Technology and Other Essays*, translated by William Lovitt (New York: Harper & Row, 1977).

Chapter 3

1. Because imagination isn't inherently moral or even wise, it can be a harmful trait in some; for example, when wielded by a clever criminal. For this reason, its status as a virtue is notoriously ambiguous.
2. Stephen Cave and Kanta Dihal's *Imagining AI* (2023) offers a rich survey of the historical origins of and differences among such narrative tropes across the globe.
3. The words of the fictional Erewhonian philosopher borrow from Butler's own prior nonfiction writings, notably his 1863 article "Darwin among the Machines."
4. Butler (1872), 199.
5. See Birhane and Van Dijk (2020).
6. See Mohamed et al. (2020).
7. Katie Booth, "What I Learned about Disability and Infanticide from Peter Singer," *Aeon*, January 10, 2018, https://aeon.co/ideas/what-i-learned-about-disability-and-infanticide-from-peter-singer.
8. The "shut up and multiply" principle has been repeatedly defended by Yudkowsky on his blog *LessWrong*, and is now invoked by many longtermists in the context of existential risks to far future people, including hypothetical digital minds. For an early discussion of its verdict in a choice between torturing one man for 50 years and putting a single dust speck in the eyes of

a vast sum of others, see Eliezer Yudkowsky, "Circular Altruism," *LessWrong*, January 22, 2008. https://www.lesswrong.com/posts/4ZzefKQwAtMo5yp99/circular-altruism. Accessed September 7, 2023.

9. Émile P. Torres, "Understanding 'longtermism': Why This Suddenly Influential Philosophy Is So Toxic," *Salon*, August 20, 2022. https://www.salon.com/2022/08/20/understanding-longtermism-why-this-suddenly-influential-philosophy-is-so/.
10. See the review of longtermist William MacAskill's stance on the Repugnant Conclusion (originally coined in 1984 by philosopher Derek Parfit) in Kieran Setiya, "The New Moral Mathematics," *Boston Review*, August 15, 2022. https://www.bostonreview.net/articles/the-new-moral-mathematics/.
11. Gideon Lewis-Kraus, "The Reluctant Prophet of Effective Altruism," *The New Yorker*, August 8, 2022. https://www.newyorker.com/magazine/2022/08/15/the-reluctant-prophet-of-effective-altruism.
12. Paris Marx, "Elon Musk's Useful Philosopher," *The New Statesman*, November 14, 2022. https://www.newstatesman.com/ideas/2022/11/elon-musk-william-macaskill-useful-philosopher.
13. Toby Ord, "Academic Bio at Future of Humanity Institute," University of Oxford, accessed September 7, 2023, https://www.fhi.ox.ac.uk/team/toby-ord/.
14. Jacob Barrett and Andreas T. Schmidt, "Longtermist Political Philosophy: An Agenda for Future Research (GPI Working Paper No. 15-2022)," Global Priorities Institute, September 2022. https://globalprioritiesinstitute.org/wp-content/uploads/Jacob-Barrett-and-Andreas-T-Schmidt-Longtermist-political-philosophy.pdf.
15. In a 2022 interview with *Vox*, MacAskill was asked why this passage was deleted from a 2021 version of the paper. He gave as a reason only that the passage might "mislead" readers, not that he had rejected the view. Sigal Samuel, "Effective Altruism's Most Controversial Idea," *Vox: Future Perfect*, September 6, 2022, https://www.vox.com/future-perfect/23298870/effective-altruism-longtermism-will-macaskill-future.
16. See Toby Ord, *The Precipice: Existential Risk and the Future of Humanity* (New York: Hachette, 2020).
17. Wiener (1961), 175–6.
18. Brynjolfsson (2022), 277.
19. Ibid.
20. Brynjolfsson (2022), 278.
21. Winner (1977), 226. See also Sullins (2013).
22. See Levy (2023).
23. John Hendel, "Crisis Text Line Ends Data-Sharing Relationship with For-Profit Spinoff," *Politico*, January 31, 2022, https://www.politico.com/news/2022/01/31/crisis-text-line-ends-data-sharing-00004001.

24. Robert Wiblin, Arden Koehler, and Keiran Harris, "Toby Ord on The Precipice and Humanity's Potential Futures," *80,000 Hours*. March 7, 2020. https://80000hours.org/podcast/episodes/toby-ord-the-precipice-existential-risk-future-humanity/.
25. Ashley Cullins and Katie Kilkenny, "As Writers Strike, AI Could Covertly Cross the Picket Line," *The Hollywood Reporter*, May 3, 2023, https://www.hollywoodreporter.com/business/business-news/writers-strike-ai-chatgpt-1235478681/.

Chapter 4

1. Whitehead (1911), 34.
2. Ibid.
3. You might think that "robot priests" would test the very limit of human tolerance for AI mirrors that parrot the human search for meaning and transcendence, but they exist nonetheless; their credibility, however, remains in question. Joshua Conrad Jackson and Kai Chi Yam, "The In-Credible Robot Priest and the Limits of Robot Workers," *Scientific American*, July 25, 2023, https://www.scientificamerican.com/article/the-in-credible-robot-priest-and-the-limits-of-robot-workers/.
4. See Sellars (1956), Section 36.
5. This example, and the chapter, are adapted from my 2021 essay "The Thoughts the Civilized Keep" in *Noema* magazine.
6. Smith (2011), 161.
7. Josh Dzieza, "AI Is a Lot of Work," *The Verge*, June 20, 2023, https://www.theverge.com/features/23764584/ai-artificial-intelligence-data-notation-labor-scale-surge-remotasks-openai-chatbots. See also Gray and Suri (2019) as well as Roberts (2019).
8. Billy Perrigo, "OpenAI Used Kenyan Workers on Less Than $2 per Hour to Make ChatGPT Less Toxic," *Time*, January 18, 2023, https://time.com/6247678/openai-chatgpt-kenya-workers/.
9. See Vallor (2015).
10. Jonas (1984), 21.
11. Jessica Riskin, "A Sort of Buzzing Inside My Head," *The New York Review*, June 25, 2023, https://www.nybooks.com/online/2023/06/25/a-sort-of-buzzing-inside-my-head/.
12. See Angwin et al. (2016).
13. See Chouldecova (2016) and Kleinberg et al. (2016).
14. Mann and O'Neil (2016).
15. James Vincent, "Automated Hiring Software Is Mistakenly Rejecting Millions of Viable Job Candidates," *The Verge*, September 6, 2021. https://www.theve

rge.com/2021/9/6/22659225/automated-hiring-software-rejecting-viable-candidates-harvard-business-school.

16. Danielle Abril, "Job Applicants are Battling AI Résumé Filters With a Hack," *Washington Post*, July 24, 2023. https://www.washingtonpost.com/technology/2023/07/24/white-font-resume-tip-keywords/?=undefined
17. Weizenbaum (1976), 38.
18. See Hartz et al. (2016).
19. Saetra (2022), 7.

Chapter 5

1. See Pierson et al. (2021).
2. "1,675 Children Removed from Parents' Custody in Benefits Scandal," *NL Times*, May 11, 2022, https://nltimes.nl/2022/05/11/1675-children-removed-parents-custody-benefits-scandal.
3. Postman (1992), 71, 118.
4. Postman (1992), 118.
5. António Guterres, "Secretary General's Opening Remarks at Press Briefing on Policy Brief on Information Integrity on Digital Platforms," United Nations, June 12, 2023, https://www.un.org/sg/en/content/sg/speeches/2023-06-12/secretary-generals-opening-remarks-press-briefing-policy-brief-information-integrity-digital-platforms.
6. Adam Minter, "Virtual Romance Is Fueling China's AI Revolution," *Bloomberg*, April 10, 2023, https://www.bloomberg.com/opinion/articles/2023-04-10/virtual-romance-is-fueling-china-s-ai-revolution?leadSource=uverify%20wall.
7. Sangeeta Singh-Kurtz, "The Man of Your Dreams," *The Cut*, March 10, 2023, https://www.thecut.com/article/ai-artificial-intelligence-chatbot-replika-boyfriend.html.
8. Samantha Cole, "'It's Hurting Like Hell': AI Companion Users Are in Crisis, Reporting Sudden Sexual Rejection," *Motherboard: Tech by Vice*, February 15, 2023, https://www.vice.com/en/article/y3py9j/ai-companion-replika-erotic-roleplay-updates.
9. Samantha Cole, "'My AI Is Sexually Harassing Me': Replika Users Say the Chatbot Has Gotten Way Too Horny," *Motherboard: Tech by Vice*, January 12, 2023, https://www.vice.com/en/article/z34d43/my-ai-is-sexually-harassing-me-replika-chatbot-nudes.
10. See Scheutz (2012).
11. Alicia Chen and Lyric Li, "'Much Less Demanding and More Manageable': Dating Is Back in China, but with an AI Twist," *Independent*, August 22, 2021, https://www.independent.co.uk/life-style/china-artificial-intelligence-dating-romance-b1900114.html.

12. See Russell (2019), Gabriel (2020), and Hadfield-Mennell et al. (2018).
13. Annie Brown, "Is Artificial Intelligence Contributing Positively to Parenting? Weighing the Pros and Cons with Angela J. Kim," *Forbes*, August 18, 2021. https://www.forbes.com/sites/anniebrown/2021/08/18/is-artificial-intelligence-contributing-positively-to-parenting-weighing-the-pros-and-cons-with-angela-j-kim/?sh=423bfc345e73.
14. Carlota Nelson, "Babies Need Humans, Not Screens," UNICEF, accessed September 7, 2023, https://www.unicef.org/parenting/child-development/babies-screen-time.
15. See Wallach and Allen (2009) as well as Wallach and Vallor (2020).
16. Sam Altman, "OpenAI CEO on GPT-4, ChatGPT, and the Future of AI," Interviewed by Lex Fridman. Lex Fridman podcast #367, March 25, 2023. https://www.youtube.com/watch?v=L_Guz73e6fw.
17. Sam Altman, "The Merge," personal blog, December 17, 2017. https://blog.samaltman.com/the-merge.
18. Sian Cain, "'This Song Sucks': Nick Cave Responds to ChatGPT Song Written in Style of Nick Cave," *The Guardian*, January 17, 2023, https://www.theguardian.com/music/2023/jan/17/this-song-sucks-nick-cave-responds-to-chatgpt-song-written-in-style-of-nick-cave.
19. See Jiang et al. (2023), 365–6.

Chapter 6

1. See Danaher and Sætra (2023).
2. See Gibson and Powys Whyte (2022).
3. Marshall McLuhan and Quentin Fiore, *The Medium Is the Massage: An Inventory of Effects* (New York: Penguin Books, 2008).
4. Le Guin (2004), 219–20.
5. Mencius (1970), 4A17; also 5A2, 6B2.
6. See Jonas (1984).
7. See Doctorow (2023).
8. See Vallor (2022).
9. See Ali et al. (2023).
10. Aristotle (1984), *Nicomachean Ethics* 1141a25–b20.
11. Arendt (1977), 193.
12. Plato (1970), *Laws* 846d.
13. Aristotle (1984), *Politics* 1278a2–12.

Chapter 7

1. See Vallor and Ganesh (2023), which incorporates research on parallels between AI governance challenges and challenges for steamboat regulation conducted by my coauthor and current PhD advisee, Bhargavi Ganesh.
2. Winner (2020), ix.
3. Weizenbaum (1976), 9.
4. Weizenbaum (1976), 225.
5. Kim Stanley Robinson, "'Mars Is Irrelevant to Us Now: We Should of Course Concentrate on Maintaining the Habitability of the Earth,'" Interview by Klaus Æ. Mogensen, *Farsight*, August 10, 2022, Copenhagen Institute for Futures Studies, https://farsight.cifs.dk/interview-kim-stanley-robinson/.
6. Mumford (1922), 170–4.
7. As a responsible disclosure, while this work was not funded by any industry engagement, I have previously received financial compensation from Google/Alphabet for expert research and training services, including employment as an AI ethicist and visiting research scientist from 2018 to 2020.
8. Winner (2020), 10.
9. Giubilini and Savulescu (2018), 170.
10. There literally was such a Church, founded by former Google and Uber engineer Anthony Levandowski in 2015. Mark Harris, "Inside the First Church of Artificial Intelligence," *Wired*, November 15, 2017. https://www.wired.com/story/anthony-levandowski-artificial-intelligence-religion/.
11. Jonas (1984), 192.
12. Jonas (1984), 162.
13. See Arista (2022).
14. Mumford (1934), 428.
15. Mumford (1934), 433.
16. Weizenbaum (1976), 32.

SELECT BIBLIOGRAPHY

Ali, Sanna J., Angèle Christin, Andrew Smart, and Riitta Katila. "Walking the Walk of AI Ethics: Organizational Challenges and the Individualization of Risk among Ethics Entrepreneurs." Paper presented at the 2023 ACM Conference on Fairness, Accountability, and Transparency (FAccT '23), Chicago, IL, USA, June 2023.

Angwin, Julia, Jeff Larson, Surya Mattu, and Lauren Kirchner. "Machine Bias." *ProPublica*, May 23, 2016. https://www.propublica.org/article/machine-bias-risk-assessments-in-criminal-sentencing

Arendt, Hannah. *Between Past and Future: Eight Exercises in Political Thought*. New York: Penguin Books, 1977.

Arista, Noelani. "Maoli Intelligence: Indigenous Data Sovereignty and Futurity." In *Imagining AI: How the World Sees Intelligent Machines*, edited by Stephen Cave and Kanta Dihal, 218–33. Oxford: Oxford University Press, 2022.

Arista, Noelani, and Sasha Costanza-Chock. *Against Reduction: Designing a Human Future with Machines*. Cambridge, MA: MIT Press, 2021.

Aristotle. *The Complete Works of Aristotle: Revised Oxford Translation*, edited by Jonathan Barnes. Princeton: Princeton University Press, 1984.

Asimov, Isaac. "Franchise." Reprinted in *Isaac Asimov: The Complete Stories*, Vol. *1*, 43–56. Broadway Books: New York, 2001.

Bender, Emily M., Timnit Gebru, Angelina McMillan-Major, and Schmargaret Schmitchell. "On the Dangers of Stochastic Parrots: Can Language Models Be Too Big?" Paper presented at the Conference on Fairness, Accountability, and Transparency (FAccT '21), Virtual Event, Canada, March 2021.

Benjamin, Ruha. *Race After Technology: Abolitionist Tools for the New Jim Code*. Cambridge: Polity Press, 2019.

Birhane, Abeba. "The Impossibility of Automating Ambiguity." *Artificial Life* 27 (2021): 44–61.

Birhane, Abeba, and Jelle Van Dijk. "Robot Rights? Let's Talk about Human Welfare Instead." Paper presented at the 2020 AAAI/ACM Conference on AI, Ethics, and Society (AIES'20). New York, February 2020.

Bostrom, Nick. *Superintelligence: Paths, Dangers, Strategies.* New York: Oxford University Press, 2014.

Bowker, Geoffrey, and Susan Leigh Star. *Sorting Things Out: Classification and Its Consequences.* Cambridge, MA: MIT Press, 1999.

Bridle, James. *Ways of Being: Animals, Plants, Machines: The Search for a Planetary Intelligence.* New York: Farrar, Straus and Giroux, 2022.

Brynjolfsson, Erik. "The Turing Trap: The Promise and Peril of Human-Like Intelligence." *Daedalus* 151, no. 2 (2022): 272–87.

Butler, Samuel. *Erewhon.* Mineola, NY: Dover Thrift Editions, 2002.

Butler, Samuel. "Darwin Among the Machines." *Letter to the Editor of* The Press, Christchurch, New Zealand, June 13, 1863.

Cave, Stephen, and Kanta Dihal, eds. *Imagining AI: How the World Sees Intelligent Machines.* Oxford: Oxford University Press, 2022.

Cave, Stephen, and Kanta Dihal. "The Whiteness of AI." *Philosophy and Technology* 33 (2020): 685–703.

Chalmers, David J. *Reality+: Virtual Worlds and the Problems of Philosophy.* New York: W.W. Norton, 2022.

Chambers, Becky. *A Closed and Common Orbit.* London: Hodder and Stoughton, 2016.

Chambers, Becky. *A Prayer for the Crown-Shy.* New York: Tor Books, 2022.

Chirimuuta, Mazviita. *The Brain Abstracted: Simplification in the History and Philosophy of Neuroscience.* Cambridge, MA: MIT Press, 2024.

Chouldecova, Alexandra. "Fair Prediction with Disparate Impact: A Study of Bias in Recidivism Prediction Instruments." *arXiv*, September 24, 2016.

Danaher, John, and Henrik Skaug Sætra. "Mechanisms of Techno-Moral Change: A Taxonomy and Overview." *Ethical Theory and Moral Practice.* 2023. https://link.springer.com/article/10.1007/s10677-023-10397-x

Descartes, René. *Meditations on First Philosophy.* 3rd ed. Translated by Donald A. Cress. Indianapolis: Hackett, 1993.

Diaz, Fernando and Michael Madaio. "Scaling Laws Do Not Scale." *arXiv:2307.03201*, July 5, 2023. https://doi.org/10.48550/arXiv.2307.03201

Dick, Philip K. *Do Androids Dream of Electric Sheep?* London: Orion Books, 1968.

Doctorow, Cory. "The Enshittification of TikTok." *Wired*, January 23, 2023. https://www.wired.com/story/tiktok-platforms-cory-doctorow

Dreyfus, Hubert. *Skillful Coping: Essays on the Phenomenology of Everyday Perception and Action*, edited by Mark A. Wrathall. New York: Oxford University Press, 2016.

Ellul, Jacques. "The Technological Order." *Technology and Culture* 3, no. 4 (1962): 394–421.

Ellul, Jacques. *The Technological Society*. New York: Knopf, 1964.

Forster, E.M. *The Machine Stops*. London: Penguin Classics, 2011.

Frankfurt, Harry. *On Bullshit*. Princeton: Princeton University Press, 2005.

Gabriel, Iason. "Artificial Intelligence, Values, and Alignment." *Minds and Machines* 30 (2020): 411–37.

Gibson, Julia D., and Kyle Powys Whyte. "Science Fiction Futures and (Re)visions of the Anthropocene." In *Oxford Handbook of Philosophy of Technology*, edited by Shannon Vallor, 473–95. New York: Oxford University Press, 2022.

Giubilini, Alberto, and Julian Savulescu. "The Artificial Moral Advisor: The 'Ideal Observer' Meets Artificial Intelligence." *Philosophy and Technology* 31 (2018): 169–88.

Godfrey-Smith, Peter. *Other Minds: The Octopus, The Sea, and the Deep Origins of Consciousness*. New York: Farrar, Straus and Giroux, 2016.

Gould, Stephen Jay. *The Mismeasure of Man*. New York: W.W. Norton. 1981.

Gray, Mary L., and Siddharth Suri. *Ghost Work: How to Stop Silicon Valley from Building a New Global Underclass*. New York: Harper Business, 2019.

Hadfield-Mennell, Dylan, Anca Dragan, Pieter Abbeel, and Stuart Russell. "Cooperative Inverse Reinforcement Learning." Paper presented at the 30th Conference on Neural Information Processing Systems, Barcelona, Spain, 2016.

Hartz, Moritz, Eric Price, and Nathan Srebro. "Equality of Opportunity in Supervised Learning." Paper presented at the 30th Conference on Neural Information Processing Systems, Barcelona, Spain, 2016.

Heinrichs, Bert, and Sebastian Knell. "Aliens in the Space of Reasons? On the Interaction Between Humans and Artificial Intelligent Agents." *Philosophy and Technology* 34 (2021): 1569–80.

Held, Virginia. *The Ethics of Care: Personal, Political and Global*. Oxford: Oxford University Press, 2006.

Hicks, Mar. *Programmed Inequality: How Britain Discarded Women Technologists and Lost Its Edge in Computing*. Cambridge, MA: MIT Press. 2018.

Jiang, Harry, Lauren Brown, Jessica Cheng, Anonymous Artist, Mehtab Khan, Abhishek Gupta, Deja Workman, Alex Hanna, Jonathan Flowers, and Timnit Gebru. "AI Art and Its Impact on Artists." Paper presented at the AAAI/ACM Conference on AI, Ethics and Society (AIES '23), Montréal, QC, Canada, August 2023.

Jonas, Hans. *The Imperative of Responsibility: In Search of an Ethics for the Technological Age*. Chicago: University of Chicago Press, 1984.

Kabir, Samia, David N. Udo-Imeh, Bonan Kou, and Tianyi Zhang. "Who Answers It Better? An In-Depth Analysis of ChatGPT and Stack Overflow Answers to Software Engineering Questions." *arXiv*, v3, August 10, 2023.

Kant, Immanuel. "What Is Enlightenment?" Translated by Ted Humphrey. Indianapolis: Hackett Publishing, 1992.

Kleinberg, Jon, Sendhil Mullainathan, and Manish Raghavan. "Inherent Trade-Offs in the Fair Determination of Risk Scores." *arXiv*, November 17, 2016.

Koopman, Colin. *How We Became Our Data: A Genealogy of the Informational Person*. Chicago: Chicago University Press, 2019.

Landgrebe, Jobst, and Barry Smith. *Why Machines Will Never Rule the World: Artificial Intelligence Without Fear*. New York: Routledge, 2022.

Leckie, Ann. *Ancillary Justice*. London: Orbit Books. 2013.

LeCun, Yann. "A Path Towards Autonomous Machine Intelligence." *OpenReview.net*, June 27, 2022. https://openreview.net/forum?id=BZ5a1r-kVsf

LeCun, Yann, and Jacob Browning. "AI and the Limits of Language." *Noema*, August 23, 2022. https://www.noemamag.com/ai-and-the-limits-of-language

Le Guin, Ursula K. "A War Without End." In *Utopia*, edited by Ursula K. Le Guin and Thomas More, 199–210. London: Verso Books, 2016.

Le Guin, Ursula K. *The Wave in the Mind: Talks and Essays on the Writer, the Reader, and the Imagination*. Boston, MA: Shambhala Publications, 2004.

Levinas, Emmanuel. *Totality and Infinity: An Essay on Exteriority*. Translated by Alphonso Lingis. Pittsburgh: Duquesne University Press, 1969.

Levy, Karen. *Data Driven: Truckers, Technology, and the New Workplace Surveillance*. Princeton: Princeton University Press, 2023.

MacIntyre, Alasdair. *Dependent Rational Animals: Why Human Beings Need the Virtues*. Chicago: Open Court, 1999.

Mann, Gideon, and Cathy O'Neil. "Hiring Algorithms Are Not Neutral." *Harvard Business Review*. December 9, 2016. https://hbr.org/2016/12/hiring-algorithms-are-not-neutral

Marcus, Gary. "GPT-5 and Irrational Exuberance." 2023. https://garymarcus.substack.com/p/gpt-5-and-irrational-exuberance

Mayor, Adrienne. *Gods and Robots: Myths, Machines, and Ancient Dreams of Technology*. Princeton: Princeton University Press, 2018.

McDowell, John. *Mind and World*. Cambridge, MA: Harvard University Press, 1999.

Mencius. *Mencius*. Translated by D.C. Lau, New York: Penguin, 1970.

Mohamed, Shakir, Marie-Therese Png, and William Isaac. "Decolonial AI: Decolonial Theory as Sociotechnical Foresight in Artificial Intelligence." *Philosophy of Technology* 33 (2020): 659–84.

Mumford, Lewis. *Technics and Civilization*. Chicago: University of Chicago Press, 2010.

Mumford, Lewis. *The Story of Utopias*. London: Azafran Books, 2017.

Noble, Safiya. *Algorithms of Oppression: How Search Engines Reinforce Racism*. New York: NYU Press, 2018.

Obermeyer, Ziad, Brian Powers, Christine Vogeli, and Sendhil Mullainathan. "Dissecting Racial Bias in an Algorithm used to Manage the Health of Populations." *Science* 366: no. 6464 (2019): 447–53.

Ortega y Gasset, José. "Man the Technician." In *History as a System and Other Essays Toward a Philosophy of History*, 87–161. Translated by Helene Weyl. New York: W.W. Norton and Company, 1961.

Ovid. *Metamorphoses: A New Verse Translation.* Translated by David Raeburn. London: Penguin Books, 2004.

Pasquale, Frank. *The Black Box Society: The Secret Algorithms that Control Money and Information.* Cambridge, MA: Harvard University Press, 2015.

Pierson, E., Cutler, D.M., Leskovec, J. et al. "An Algorithmic Approach to Reducing Unexplained Pain Disparities in Underserved Populations." *Nature Medicine* 27 (2021): 136–40.

Plath, Sylvia. "Mirror." Reprinted in *Crossing the Water.* New York: Harper and Row, 1971.

Plato, *The Laws.* Translated by Trevor J. Saunders. London: Penguin Books, 1970.

Popper, Karl. *The Open Society and Its Enemies.* New York: Routledge, 2002.

Postman, Neil. *Technopoly: The Surrender of Culture to Technology.* New York: Knopf, 1992.

Roberts, Sarah T. *Behind the Screen: Content Moderation in the Shadows of Social Media.* New Haven, CT: Yale University Press, 2019.

Robinson, Kim Stanley. *The Ministry for the Future.* London: Orbit Books, 2020.

Russell, Stuart J. *Human Compatible: Artificial Intelligence and the Problem of Control.* New York: Viking, 2019.

Sætra, Henrik Skaug. "Robotomorphy: Becoming Our Creations." *AI and Ethics* 2 (2022), 5–13.

Scheutz, Matthias. "The Inherent Dangers of Unidirectional Emotional Bonds Between Humans and Social Robots." In *Robot Ethics,* edited by Patrick Lin, Keith Abney, and George Bekey, 205–22. Cambridge, MA: MIT Press, 2012.

Sellars, Wilfrid. *Empiricism and the Philosophy of Mind,* edited by Robert Brandom. Cambridge, MA: Harvard University Press, 1997.

Smith, Benedict. *Particularism and the Space of Moral Reasons.* New York: Palgrave Macmillan, 2011.

Stephenson, Neal. 1995. *The Diamond Age: Or, A Young Lady's Illustrated Primer.* New York: Bantam Books.

Sullins, John P. "An Ethical Analysis of the Case for Robotic Weapons Arms Control." Paper presented at the 5th International Conference on Cyber Conflict, Tallinn, Estonia, 2013.

Tronto, Joan. *Moral Boundaries: A Political Argument for an Ethic of Care.* New York: Routledge, 1993.

Turing, Alan M. "Computing Machinery and Intelligence." *Mind* 59 (1950), 433–60.

Turkle, Sherry. *Alone Together: Why We Expect More from Technology and Less from Each Other.* New York: Basic Books, 2011.

Vallor, Shannon. "The Fantasy of Third-Person Science: Phenomenology, Ontology and Evidence." *Phenomenology and the Cognitive Sciences* 8, no. 1 (2009): 1–15.

Vallor, Shannon. "Moral Deskilling and Upskilling in a New Machine Age: Reflections on the Ambiguous Future of Character," *Philosophy and Technology* 28 (2015), 107–24.

Vallor, Shannon. *Technology and the Virtues: A Philosophical Guide to a Future Worth Wanting*. New York: Oxford University Press, 2016.

Vallor, Shannon. "The Thoughts the Civilized Keep." *Noema*. February 2, 2021. https://www.noemamag.com/the-thoughts-the-civilized-keep

Vallor, Shannon. "We Used to Get Excited about Technology: What Happened?" *MIT Tech Review*. October 21, 2022. https://www.technologyreview.com/2022/10/21/1061260/innovation-technology-what-happened

Vallor, Shannon and Bhargavi Ganesh. "Artificial Intelligence and the Imperative of Responsibility: Reconceiving AI Governance as Social Care." In *The Routledge Handbook of Philosophy of Responsibility*, edited by Maximilian Kiener, 395–406. New York: Routledge, 2023.

Verbeek, Peter-Paul. *Moralizing Technology: Understanding and Designing the Morality of Things*. Chicago: University of Chicago Press, 2011.

Verbeek, Peter-Paul. *What Things Do: Philosophical Reflections on Technology, Agency and Design*. University Park: Pennsylvania State University Press, 2005.

Vinsel, Lee, and Andrew L. Russell. *The Innovation Delusion: How Our Obsession with the New Has Disrupted the Work That Matters Most*. New York: Currency, 2020.

Wallach, Wendell, and Colin Allen. *Moral Machines: Teaching Robots Right from Wrong*. New York: Oxford University Press, 2009.

Wallach, Wendell, and Shannon Vallor. "Moral Machines: From Value Alignment to Embodied Virtue." In *Ethics of Artificial Intelligence*, edited by Matthew Liao, 383–412. New York: Oxford University Press, 2020.

Wang, Boshi, Xiang Yue and Huan Sun. "Can ChatGPT Defend its Belief in Truth? Evaluating LLM Reasoning via Debate." *Findings of the Association for Computational Linguistics: EMNLP 2023*, 11865–11881, Dec 6–10, 2023. Association for Computational Linguistics. https://aclanthology.org/2023.findings-emnlp.795.pdf

Weizenbaum, Joseph. *Computer Power and Human Reason*. New York: W.H. Freeman, 1976.

Wells, Martha. *The Complete Murderbot Diaries*. New York: Tor.com, 2018.

Whitehead, Alfred North. *Introduction to Mathematics*. New York: Henry Holt, 1911.

Wiener, Norbert. *Cybernetics, or Control and Communication in the Animal and the Machine*. 2nd ed. Cambridge, MA: MIT Press, 2013.

Winner, Langdon. *Autonomous Technology*. Cambridge, MA: MIT Press, 1977.

Winner, Langdon. *The Whale and the Reactor: A Search for Limits in an Age of High Technology*. 2nd ed. Chicago: University of Chicago Press, 2020.

INDEX

For the benefit of digital users, indexed terms that span two pages (e.g., 52–53) may, on occasion, appear on only one of those pages.

Notes are indicated by "n" following the page number.

accountability, 52–53, 62, 119, 127–28, 175, 196. *See also* answerability
accuracy, 24–25, 51–52. *See also* truth
advisors, artificial, 29–30, 36, 54, 115–16, 152, 212–13, 217
affordances, technological, 140–41
agency, machine versus human, 72, 74, 89–90, 138–39, 157–58, 178
agents, artificial. *See* chatbots
AGI. *See* artificial general intelligence
AI
 conservativism of, 57–58, 224–25
 ethics, 45, 62, 173, 187, 196
 industry leaders, 86, 96, 152, 198–99
 reforms of, 45, 62, 129–30, 170, 196
 responsible, 62, 173, 213–14
 safety, 9–10, 76–77, 81–83, 95, 169–70, 196. *See also* value alignment
algorithms
 applications of, 3, 41, 115, 122–26, 137, 211–12
 definition and origin of, 104
 for cognitive automation, 102–3, 110, 112
 social power of, 104
 See also bias; machine learning
alignment. *See* value alignment
Altman, Sam, 75, 86, 152
androids, 21–22, 70, 153
answerability, 121–22, 127. *See also* accountability
anthropocentrism, 17, 83
anti-humanism, 155–56
apocalypse, 152, 164. *See also* existential risk
Aristotle, 66, 68–69, 100, 167, 169, 177, 190–91, 203, 208

art, AI-generated, 29, 99, 141–42, 157–60, 200–1
artificial general intelligence (AGI)
  definition and characteristics of, 21–22, 38–39, 206
  extermination of humanity by, 4, 82–83, 95, 152–53, 196
  narratives, 69, 71, 91–96, 217
  path toward, 22–24, 28–29, 35, 90
  predicted dangers from, 71, 73–74, 76–77, 80–85
  See also existential risk; superintelligence; superhuman AI
arts
  and artists, 157–59, 209
  humane, 171–72, 209–10
  mechanical and technical, 171–72, 190–92, 208, 209–10
  See also creative work; creativity; domestic skill and expertise; technê
augmentation of human capabilities, 87–88, 98–99, 130, 189
authoritarianism, 109, 155–56, 164, 185, 215
autofabrication
  definition and examples of, 12, 158–59, 206
  engines of, 101, 201–2
  as existential task, 67, 90–91, 111–12, 156–57
  and practical wisdom, 118–19, 193
automata, 16–17, 142, 230n.2
automated decision-making
  applications of, 101, 103, 106, 115, 122, 125, 170–71
  critical questions for developers about, 130
  and discrimination, 110–11
  human oversight of, 116–17, 212
automation of labor, 66–67, 87, 103, 110, 139, 215. *See also* future of work; labor
autonomous machines, 15, 61, 196, 204–5. *See also* self-driving cars
awareness, 23–24, 26, 54, 58–59. *See also* consciousness; experience; minds; sentience

behavior
  manipulation and nudging by AI, 86–87, 138, 163, 175
  analysis and prediction of human, 58, 83, 89
  unpredictability of machine, 104–5, 196
Benjamin, Ruha, 44–45, 75, 211
bias
  algorithmic, 41–46, 122–23, 125–27, 129, 136–37, 210–11
  in humans, 43–45, 46, 133–37, 152, 182–83, 199–200, 211
  learning or inductive, 42
  See also discrimination; machine learning fairness; runaway feedback loops
biological
  bootloader for AI, 152
  discoveries, 28, 215
  evolution of intelligence, 71
  inspiration in AI research, 17, 83
Birhane, Abeba, 57, 74
Blade Runner, 21–22, 153–55
bootstrapping, moral, 163–64, 167, 169–70, 178, 181–82, 209–10
brain, human, 24, 26, 28–29, 38–40, 102–3
brain-computer interfaces, 153–54, 178
Bridle, James, 83
Brynjolfsson, Erik, 84, 87–90
bullshit
  definition of, 119–20
  AI-generated, 120–22, 145–46, 157–58, 184, 214
  See also fabrication

Butler, Samuel, 70–75, 85–86, 91, 94–95, 190. *See also Erewhon*

calculus, moral, 78, 98, 157, 169–70. *See also* longtermism; utilitarianism
capabilities
of AI systems, 22–23, 29–30, 51, 72, 86, 196, 229n.3
cognitive, 8, 12, 29, 30, 86, 103
creative, 139, 158, 168–69
empathic and emotional, 91–92, 146–47, 148, 149–50
moral and political, 7–8, 117, 122, 138–39, 153–54, 156–57, 168–69
scientific and technological, 8–9, 38–39, 178, 202–3, 219
severance of humane and technical, 166, 202–3, 209–10
capitalism, 166–67, 179, 196–97
care
domestic, 190–91, 208
ethical duties of, 170, 176, 189, 198, 206
human capacity for, 142–43, 156, 190–92, 203, 210–11, 216–17, 220–22
social, 56, 63–64, 110, 192
See also healthcare
Cave, Nick, 157–59, 201
Chambers, Becky, 220–21
character, moral, 7–8, 48, 165–69, 162–63, 169–70, 181–82, 185. *See also* virtues
chatbots
for companionship, 143–51, 154
dangers of, 37–38, 56, 87
therapeutic and advisory uses of, 55–56, 87, 188, 200–1, 213
ChatGPT, 19–20, 25–26, 30–31, 51–52, 120–21, 157–58, 184, 214. *See also* chatbots
civil rights, 110–11, 135, 151. *See also* human rights
classification, 19–20, 22, 27–28, 42, 50, 182
climate change, 7, 78–80, 88, 95–98, 156, 165, 170–71
See also environmental applications of AI; environmental ruin
cognition, theories of human and machine, 17, 104–5, 189
cognitive
energy and faculties, 12, 18, 24, 30, 103
limitations and imperfections of humans, 6, 113, 189
tasks and labor, 21, 22–23, 103
collective
action, 95, 98, 111–12, 117, 157, 208–9, 219
capacities, 4, 8, 88, 138–39, 157
will and judgment, 3, 88, 221–22
wisdom, 111–12, 118, 165, 169, 193
colonial legacy of AI, 18, 74–75, 91
common-sense world model, 26, 31, 84
compassion, 91–92, 97–98, 201–2
computational theory of mind, 40–41, 69–70
Confucian moral thought, 66, 168–69, 203–4
consciousness
embodied nature of, 24, 231n.4
evidential and intersubjective reality of, 58–59
machine, 1, 70, 72, 92, 155, 206, 219, 231n.4
See also awareness; experience; minds; sentience
contestability of AI decisions, 110–11, 123, 177–78, 196, 215–16
cost-benefit analysis, 149

courage
civil, 118–19, 219–20, 224
moral virtue of, 65–69, 97–99, 101
crafts. *See* arts, mechanical and technical
creative work, 139–42, 157–58, 215. *See also* arts; generative AI and employment
creativity
expressed through technology, 11, 192, 206, 225
human, 7, 11, 29–30, 99, 139–42, 157–59
and imagination, 65–66, 97–98, 166
and self-renewal by autofabrication, 156–59, 193, 219–20
machine, 141–42
moral, 97–98, 157–58, 164–65, 168–69, 178
See also arts; expression; novelty
cultural record, 133–34
cybernetics, 17, 216–17

data science, 20–21
data sets. *See* training data
data structures, 26, 28. *See also* patterns, statistical
deception, 2–3, 119–20, 143–47. *See also* manipulation
decision-making, human, 87, 108–11, 116–17, 119, 121, 129–31. *See also* automated decision-making
deepfakes, 115, 143
deep learning, 29, 105–6, 115, 119, 127–28
deliberation, 85, 117, 123, 163
deliverance, 186
democracy, 111–13, 131, 143, 170, 176–77, 184, 215
democratic
improvisation, 165
legitimization of power, 175
values, norms and ideals, 104–5, 172, 199, 202, 211–12
dependence
human, 191, 206–7, 223
on technology, 6, 73, 104, 146, 176–77
See also interdependence
deregulation, 175, 196–97
Descartes, René, 39–40
deskilling, moral, 117
devaluation of humans, 134, 139, 142, 190, 230n.1
Dick, Philip K., 153–57, 207
digital people, 78, 85, 155, 169–70
disability, 45, 77–78, 126, 210–11
discrimination, 41–44, 110–11, 126–27, 135, 173. *See also* bias; machine learning fairness
disempowerment, human, 74, 87, 124–25
disinformation, 115, 143, 170–71
disposability, 188–89, 217
divine AI creation, 218–19, *See also* gods; theology of AI
domestic skill, expertise, and virtues, 8, 190, 191, 203, 208, 221
See also arts, mechanical and technical; care; service; technê
driverless vehicles. *See* self-driving cars
dystopian visions, 94–95. *See also* apocalypse; existential risk

economic order
impact of AI on, 2, 15
incentives of, 9, 89–90, 99, 139–42, 196–97, 204, 210
inequality and marginalization in, 55, 133–34, 162, 199
productivity and utility of AI in, 85–87, 131, 140–41, 192
replacement and disempowerment of humans in, 72, 87, 205

sustainability of, 188–89, 196–97, 216
values and virtues of, 9–10, 141–42, 148, 166–67, 180–82, 202
education
engineering ethics in, 174–76
modern public, 131
necessary reforms in systems of, 166, 179–80
split between humane and technical arts in, 171–72, 204
uses of AI in, 106, 214–15, 222–23
effective altruism, 76–79, 157, 233n.15. *See also* longtermism
efficiency
of AI solutions, 26, 61–62, 81–82, 136–37, 166, 200–1
contrast with human behavior, 11, 86, 116–17, 129, 131–32, 150–53
rule and values of, 186–87, 200–201, 217
tension with humane virtues, 8, 11, 89–90, 186–87
See also optimization
Ellul, Jacques, 142–43, 186–87
embodiment
of AI, 70, 206, 219, 231n.4
of biological minds and intelligence, 40–41, 83, 220–21, 230n.6, 231n.2
of consciousness and meaning, 24, 33–34
of human knowledge and experience, 6, 33–34, 48, 205
emotion
absence in AI of, 141–43, 145–46, 152
AI manipulation of, 146–48
in AI language use and design, 37–38, 143–44
and human decision-making, 124, 129–30
monitoring and prediction of human, 89, 146–47, 153–54
empathy
AI facsimiles and artificial stimulation of, 89–90, 147, 153–54
varieties of, 146–47
virtue of, 69, 156–57
employment. *See* hiring applications of AI; future of work; labor
engineering
of automata and intelligent systems, 16–19, 45, 105–6, 138
culture and professional societies of, 170, 174–75, 197
ethical codes and education in, 170, 174–76
of future selves and communities, 31, 131–32, 164
and gender bias, 41, 125–26
and human nature, 12–13
increased ethical demands on profession of, 170–71, 176–78
social power and responsibility of, 173–76
Enlightenment, 109–10, 131, 183
environmental
applications of AI, 8–9, 28, 135
costs and footprint of AI, 9, 215–16
ruin, 37, 80, 111–12, 167, 198–99
See also climate change; existential risk; sustainable futures
epiphany, 56–57
epistemic caution, 162
equitable
futures, 159–60, 172, 196, 210–11, 214–15, 220
outcomes of AI decisions, 123, 129–30
Erewhon, 70–75, 190. *See also* Butler, Samuel
error in AI systems, 25–26, 45

ethical theory. *See* moral philosophy
eugenics, 18
evolutionary metaphor for AI development, 71, 73, 152
excellence. *See* virtues
existentialist philosophy, 12, 195–96, 229n.4
existential risk
  from AGI, 6–7, 77, 80–82, 94–96, 161–62
  of civilizational collapse, 161–62, 164
  from climate change, 7, 98
  from human incapacity, 7, 195–96
  to far future people, 65, 77, 232–33n.8
  in science fiction narratives, 76, 91, 98, 165
  See also climate change, longtermism
experience, lived
  absence in AI of, 23–24, 63–64, 141–42, 158, 194
  in learning and expertise, 26–27, 89, 100–1, 117–18
  human, 24–25, 58–59, 94–95, 142, 153–54, 158–59, 191, 205
  of others, 48, 59–61, 142–48
  physical character of, 145–46
  See also awareness; consciousness; sentience
expertise, 166, 171–72, 176–78, 190, 191
explainable AI, 62, 106. *See also* interpretable AI models; opacity in AI
expression, 141–42, 158–60, 180, 192, 194, 201–2. *See also* creativity
extinction
  human, 6–7, 71, 80, 96, 161–62, 198–99
  sixth mass, 9–10, 98, 165
extraction
  of resources, 75, 118, 166–67, 179, 190–91, 210, 221–22
  in human relationships, 150–51
  of patterns from data, 10–11, 20–21, 24–25, 38, 41, 89, 101, 105

fabrication by generative AI models, 24–26, 31, 51–52, 80–81, 120–21, 214. *See also* accuracy; bullshit
facial tracking and recognition, 22, 60–61, 89, 126
fairness. *See* machine learning fairness
faith, 138, 179, 195–96, 200, 217, 218, 229n.4
  See also theology of AI
fantasies
  about AGI, 78–79, 94–96, 152–53
  of digital salvation, 40–41, 70, 215, 218–19
  and imagination, 65, 68, 76
filters, algorithmic, 41–42, 44–45, 54–55, 89, 125, 137–38, 186–87
financial applications of AI, 15, 135, 211
fine-tuning of AI models, 22–23, 28–31, 121–22, 139–40
flexibility of human intelligence, 85–86, 168–69
flourishing, human and planetary, 8–9, 67, 111–12, 156–57, 161–62, 171, 176–79, 207
Franchise, 112–15
Frankfurt, Harry, 119–20
fraud detection algorithms, 106, 137, 211
friction, human-generated, 186, 188–89
future of work, 4, 85, 215. *See also* generative AI and employment; labor

Gasset, José Ortega y, 12–13, 67, 204–5, 229n.4. *See also* autofabrication

Gebru, Timnit, 32–33, 173
generative AI
and culture, 3, 133–34
and fabrication of falsehoods, 24–26, 51–52, 80–81, 111
and employment, 4, 29–30, 99, 139–44
failures of understanding in, 30–31
novelty produced by, 141–42, 159
uses of, 29–30, 49–50, 214–15
See also large language models; image models; multimodal AI
generosity, 65–66, 159, 225
Global North, 48–49, 77
gods, machine, 51, 54, 217. *See also* divine AI creation; theology of AI
GOFAI (good old-fashioned AI), 19–21
governance of technology, 174, 196–97, 215–16, 221–22
government uses of AI and algorithms, 55–56, 137, 172–73, 210–11
GPT. *See* large language models

habits, moral and intellectual, 69, 97–98, 102, 161–63, 166–67, 169–70, 181–83
habituation, 66–67
hallucination. *See* fabrication by AI models
happiness, utilitarian calculus of, 77–78
harms from AI, 4, 41–42, 123, 136–37, 182, 196–97
healing, 4, 208, 222
healthcare, AI in, 28, 42–44, 56, 106, 115, 135–37, 177–78, 210–11
hierarchies, 71, 74, 75, 169, 182–83, 192, 203
hiring applications of AI, 41, 50–51, 125–28
HLAI (human-like artificial intelligence), 84, 85, 88–89
homogeneity in contemporary technological culture and values, 13, 133–34, 180
honesty, 65–66, 69, 177
humane
feeling, 78, 151–52
futures, 7, 202, 210, 225
ideas, values and visions, 202–4, 218
knowledge and wisdom, 11, 55–56, 209–10
potential and possibilities, 10–11, 13–14, 89–90, 166, 202–3, 219, 224
virtues, 8, 65–67
See also arts, humane
human-in-the-loop. *See* oversight, human
humanities, 171–72, 209
human rights, 77, 109–10, 143, 151, 196, 207, 211–12. *See also* legal rights and protections

ideology, 109, 218
image models, 22–23, 53–54, 99, 137–38
imagination
in children and fantasy, 65–66, 68–69, 76, 97–98
creative power of, 76, 98–99, 128–29, 156–57
limitations and failures of, 96–98, 232n.1
moral, 96–99, 101, 111, 117–18, 128–29, 166
virtue of, 65–67, 98, 101, 117–18, 156–57
See also fantasies
immortality, digital, 40–41, 70
imperfection, human, 45, 129, 151–52, 155, 202, 219, 223
incentives, 84–86, 136–37, 140, 205, 210–11. *See also* economic order, incentives of

inclusive AI development,
62–63, 139–40
independent thinking, 167, 183–84
Indigenous peoples and communities,
165, 218, 222–23
industrial and postindustrial threats,
118, 173–74, 208–9
inequality. *See* economic order, inequality and marginalization in
inevitability, false narratives of, 8, 72,
75, 94–95, 111–12
inferiority, 73, 74, 190–91
information, 8, 42–43, 50–51,
124, 217
injustice, 122–23, 134–37, 182–83,
200–1, 210–11
innovation
critiques of, 88, 188–89, 200–1,
208–9
culture of, 172, 181–82
incentives of, 215–16, 224
meaning and necessity of,
189, 198–99
moral, social, and political, 117–
18, 224–25
and regulation, 197–98
intellectual virtues, 65–67, 183. *See also* practical wisdom
intelligence
biological and embodied, 24, 40–41,
83–84, 89–90, 94, 230n.6, 231n.2
presumed malicious character of,
82–83, 93, 95
moral and social, 54–55, 68, 168–70,
178, 224
superhuman, 90, 93–94, 218
testing, 18
See also artificial general intelligence; HLAI; minds; superintelligence
interdependence, human, 156, 161–62,
192, 202. *See also* dependence
interpretable AI models, 106, 122, 177–78,
188–89. *See also* uninterpretability

jobs. *See* future of work
Jonas, Hans, 118–19, 171, 204–
5, 218–19
judicial use of AI systems, 104, 122,
136–37, 211
justice
human capacity for, 63–64, 89–90
demands of, 60–63, 170, 177–78,
202–4, 207
virtue of, 67, 99
justification, moral and intellectual,
106–7, 109, 116, 130

knowledge, 8, 61, 87, 100–1, 106–9,
205, 222–23
See also humane knowledge and wisdom, self-knowledge; technê
kubernetes, 216–17. *See also* cybernetics

labor
markets, 85, 125
needed to maintain AI, 81, 114
organizing, 99, 188
precarious or unpaid, 74, 114,
187, 200–1
See also automation; creative work; future of work; generative AI and employment
large language models (LLMs)
and AGI, 84
capacities of, 1–2, 22–23, 49–50,
115–16, 142, 152
design and workings of, 32–35, 105
limitations and risks of, 31, 32, 56,
61–63, 107–8, 120–21
non-neutrality of, 133–35
See also ChatGPT, generative AI, machine learning

law enforcement uses of AI, 57, 115, 135, 211
lawmaking, 159–60, 197, 213–14
layers and weights in AI models, 105–6. *See also* deep learning
leadership, 109, 126, 167, 173, 184, 199, 203, 215–16
learning
  community language, 223
  from experience, 100
  moral, 8, 66, 128–29, 134–35
  unsupervised, 19–20, 84, 115
  See also deep learning; machine learning, reinforcement learning
LeCun, Yann, 29–30, 33, 84
legal rights and protections, 110–12, 151–52, 196, 207. *See also* human rights
Le Guin, Ursula, 166, 202
Lemoine, Blake, 1–3
Levinas, Emmanuel, 60–61
LLMs. *See* large language models
loneliness, 144
longtermism, 76–80, 94–98, 155, 157, 169–70, 217, 232–33n.8. *See also* effective altruism
love, 48, 145–49, 157, 159–60, 192, 200–1, 203
lying, 119–21

MacAskill, William, 78–80, 233n.15
machine ethics, 152, 185–86, 217
machine learning
  advantages over human cognition, 26–28
  dangers and risks of, 81–83, 104–5, 107–8
  fairness, 42, 62, 129, 177–78, 196
  methods, techniques and varieties of, 18–20, 22, 35, 47–50, 105–6
  mirror properties of, 47–50
  See also deep learning; generative AI; large language models
MacIntyre, Alasdair, 191, 206
magnifying power of AI mirrors, 13–14, 105, 122–23, 125, 136–37, 213
malevolence in AGI, 80–82, 95–96
manipulation, 80, 86–87, 143–44, 146–47
Marcus, Gary, 30–31
marginalized communities and groups, 46, 74–75, 129, 133–34, 173, 211–12
mathematical nature of AI tools, 18–19, 26, 30, 38–39, 41, 61, 108, 149, 158
meaning, 31, 53, 86–87, 141–42, 201, 222–23
mechanization of the human personality, 11, 89–90, 142–43. *See also* reverse adaptation
medical uses of AI. *See* healthcare
mental health, 44–45, 200–1. *See also* loneliness
metaphor, 40–42, 229n.1
mimicry, 17, 26, 29–30, 38, 84–86, 122, 158
minds, 39–41, 61, 69–70, 194–95, 231n.2, 231n.4. *See also* awareness; consciousness; intelligence; sentience
mirrors, properties of, 38–39, 46–47, 49–50, 62, 100, 138
misinformation, 46, 143
moral
  experience, 109, 153–54
  machines (*see* machine ethics)
  obligations, 61, 78–79, 207
  philosophy, 77–79, 178, 203
  reasoning, 108–12, 117–18, 128, 177
  responsibility, 13, 63, 111, 129, 166, 174–75, 196

moral (*cont.*)
See also knowledge, moral; innovation, moral and political; space of reasons; virtues
mortality, 40–41, 135, 172, 199
multimodal AI, 22–23
Multivac, 112–14, 116, 124–25, 131
Murderbot, 219–20

Narcissus and Echo, 4–8, 10–11, 15, 34–35, 46, 64, 143–44, 219
narrow AI, 22–23, 85–88
neural networks, artificial, 26
neutrality, 12–13, 45, 109, 131, 134, 166, 202
noise, statistical, 48–49, 141–42, 159
normative judgment, 158–59, 198–99
novelty, 19–20, 29, 32–33, 140–41, 159, 188–89
nuclear weapons, 6–7, 78–80, 82–83, 118

objective function, 18–19, 49, 53, 81–82, 95, 205
objectivity, 133–34, 167
occlusions of AI mirrors, 13–14, 47, 52, 56–57, 58, 62, 114, 142–43, 155–56
oligarchy, 131, 199
opacity in AI, 104–6, 110–11, 113–15, 119, 122, 127–28
open-source models, 37–38
oppression, 74–75, 134–35, 185, 195–96
optimization
in AGI, 86–87, 91–92, 206
of human behavior, 11, 37–38, 61, 88–90, 148–49, 152–53
instrumental values and policies of, 77–78, 200–1, 217, 221–22
in machine learning algorithms, 18–19, 22, 31, 38, 49, 105–6, 140
Ord, Toby, 79–80, 94, 96–97
oversight, human, 27–28, 212

pain, 38, 146–47, 154–55, 159, 194
parameters, 49, 81, 105. *See also* weights
patterns
breaking of, 158–60
creative, 140–41, 158–59
discriminatory, 42–44
historical, 10–11, 58, 75, 100–1, 133–35, 136–37, 200–1
in language, 29, 32–33, 38–39, 120–21
of moral thought, 128–29, 162–63, 166–69, 178, 181–82
statistical, 90–91, 105
unsustainable, 31, 183
See also extraction of patterns from data
perseverance, 167, 184–85
personality
algorithmic, 50
chatbot, 144–45, 148
human 11, 12, 138–39, 142–43, 207, 230n.1
scientific, 209
phenomenology. *See* experience, lived
philanthropy, 76–78, 80, 95–96
phrónēsis. *See* practical wisdom
pilot, 40, 216–19. *See also kubernetes*
planetary ruin. *See* environmental ruin
platform technology companies, 20, 48–49, 53, 85, 172, 175, 184–85, 190–91
Plato, 175, 190, 202–3
play, 91–94, 159, 192, 207
polarization, 46, 55, 138, 199
policing uses of AI. *See* law enforcement uses of AI
policy, 69, 76–77, 79–80, 91, 94, 103, 111, 213–14

political
  action, 111–12, 188, 203, 224
  authority, 175
  imagination and will, 90–91, 99, 196–97
  incentives, 89–90, 210
  influence, 79–80, 87
  innovation, 117–18, 151, 164, 224
  wisdom, 138, 166, 169, 176–77, 179
polypotency of AI, 140
possibility, human, 10–11, 13–14, 58, 97–98, 118–19, 166, 193, 219
Postman, Neil, 142–43
potential, human, 11, 13, 56–57, 63, 66–67, 89–90, 166, 202–3
power
  abuse and corruption of, 95, 173–74, 215–16
  concentration of, 75, 87, 157
  humane and creative, 7–8, 11, 61, 65–66, 89–90, 99, 159–60
  legitimacy of, 173–75
  and responsibility, 31, 118
practical wisdom, 68–69, 99–100, 117–19, 164–65, 168–69, 177. *See also* prudence
prediction, 10, 28, 38–39, 49–50, 90–91, 133–34, 200–1. *See also* machine learning
productivity, 85–87, 150–51, 166–67, 187, 200
professional responsibilities and virtues, 170, 174–77
progress, 11, 103, 117, 198–99
proprietary technology, 20, 104–5, 122–23, 126–28
proxies in data and algorithms, 41–43, 123
prudence, 68, 99–101. *See also* practical wisdom
racial and gender bias in AI. *See* bias
rationality, instrumental values and, 148–50, 221
reasoning. *See* space of reasons
reciprocity, 143–44, 148
recommendations, algorithmic, 3, 21–22, 49–50, 110–11, 116–17, 177–78
redemption, 56–57
redress, 110–11, 196
refusal and relinquishment, 64, 185–86, 188–89, 201–2, 218–19
regulation, 6–7, 82–83, 130, 174–75, 196–98
reinforcement learning, 19–20, 29, 53–54, 84, 149–50. *See also* machine learning
repair, 188–89, 221–23, 225
replacement, human, 63–64, 71–72, 76, 145, 149, 151–52
resistance, 109–12, 131–32, 186, 188–89
responsibility. *See* moral responsibility
restoration, 188–89, 206, 210, 221–22. *See also* repair
restraint, 220–22
revalorization, 188–89
reverse adaptation, 88–91
revolution, 172, 215–16, 225
reward hacking, 81–83. *See also* objective function
robotics, 16–17, 20–21, 23–24, 147
robots, 16–17, 20–21, 23–24, 40–41, 44–45, 70, 71, 88–89
runaway feedback loops, 44–45, 57, 137–38, 211

Sætra, Henrik Skaug, 129–30, 230n.1
science fiction, 71–73, 76, 112–15, 153–55, 166, 219–22
self-fulfilling prophecies, 10, 101, 138
self-determination, 3, 138–39

self-driving cars, 23–24, 26–27, 61, 215
self-forgetting, 2, 8, 150, 152–53, 57, 202–3
self-knowledge, 2, 3, 6, 55–56, 153–54. *See also* self-understanding
self-renewal, 56–57, 156–57, 179–80, 193
self-understanding, 31, 47, 63, 163, 195–96. *See also* self-knowledge
sentience, 1, 17, 92, 146–47, 153–54, 206, 207, 219. *See also* awareness; consciousness; experience
service, 62–63, 189–91, 192, 203, 208–10, 216–17, 225
skillful coping and performance, 24–27, 89, 107, 203–4, 208–9
sociopaths, 54, 109, 146–47
solidarity, 8, 48, 61, 62–64, 91–92, 118–19, 192, 212
space of reasons, 106–10, 116, 121, 123–24, 128, 129–32
speed, 22–23, 88–89, 104–5, 186, 188–89, 217
spontaneity, 56–57, 90
standards, 45, 62, 82–83, 122, 170, 177, 196, 198–99
stereotypes, 51, 53–54
stochastic parrots, 32–33
structural change, 182–83, 224–25
suffering, 77–78, 98, 154–55, 216. *See also* pain; sentience
superhuman AI, 72–74, 86–87, 90–94, 218–19. *See also* artificial general intelligence; superintelligence
superintelligence, 80, 94, 217. *See also* artificial general intelligence; superhuman AI
supremacy, 73, 75–76, 91–92
surveillance, 182–83
survival, human, 7, 63, 97, 118–19, 156–57, 164–65, 170–71, 195–96
sustainable futures, 8–9, 178–79, 188–89, 192, 196–97, 202, 215–16, 221–23

techlash, 88, 172
technê, 191. *See also* arts, mechanical and technical; knowledge, domestic and mechanical
technology, heart of, 189, 191, 225
technomoral change, 164
technopessimism, 188
technopoly, 142–43
techno-utopians, 207–8, 215, 223, 225. *See also* theology of AI; utopian ideologies
theology of AI, 217–19. *See also* techno-utopians
therapeutic uses of AI, 37, 84, 87–88, 145, 188
tokens, 38, 142, 195–96
training data, 19–20, 42–44, 48–49, 52, 62, 80–81, 137–38, 186–87
transparency, 104–5, 110–11, 122, 124–25, 129–30, 138–39, 196, 215–16
trust and confidence in science and technology, 88, 171–75
truth, 23–25, 46, 119–21, 133–34, 145–46, 200–1. *See also* accuracy
Turing Test, 18
Turing Trap, 87–89

understanding, 15–16, 30–32. *See also* self-understanding
uninterpretability, 104–6. *See also* opacity in AI
unmaking, 158–60, 195
unpredictability, 26–27, 104–5, 196–98
unsustainable patterns and practices, 31, 136–37, 166–68, 183

uploading, 40–41, 70, 93, 144–45
utilitarianism, 77–79, 95, 169–70
utopian ideologies, 165, 169–70, 207–8, 218–19, 223–25. *See also* techno-utopians

value alignment, 149–50, 187, 229n.3. *See also* AI safety
values
  dominant, 9–10, 133–34, 139–42, 156–57, 162–63, 187, 202
  domestic, 221–22
  technical and instrumental, 186–87, 217
  woven in technology 12–13, 35, 133, 201–2
vices, 48, 68–69, 166–67, 185–86
virtual worlds, 40–41, 78, 93, 169–70, 207
virtues
  cultivation of, 66, 98, 117–19, 148, 183
  definition and study of, 65–67, 169, 203
  of the modern industrial order, 166–67
  technomoral, 179
  See also care; courage; empathy; imagination; love; practical wisdom; repair; service
vital pretechnical program, 205–206, 208, 219, 221–22
vulnerability, human, 191–92, 206

weights, 18–19, 105–6, 113–14. *See also* parameters
Weizenbaum, Joseph, 129, 204–5, 217, 219, 224–25
Westworld, 76, 91–95
Whitehead, Alfred North, 102–3, 117
Wiener, Norbert, 81–82, 216–17
Winner, Langdon, 88–89, 199, 216–17
wisdom, 8, 66–69, 100–1, 124, 138, 169, 171, 204–5. *See also* practical wisdom
world-model, 26, 84

zero-sum games, 91–92, 101